Rad-hard Semiconductor Memories

RIVER PUBLISHERS SERIES IN ELECTRONIC MATERIALS AND DEVICES

Series Editors:

Edoardo Charbon
EPFL, Switzerland

Mikael Östling
KTH Stockholm, Sweden

Albert Wang
University of California, Riverside, USA

Indexing: All books published in this series are submitted to the Web of Science Book Citation Index (BkCI), to CrossRef and to Google Scholar.

The "River Publishers Series in Electronic Materials and Devices" is a series of comprehensive academic and professional books which focus on the theory and applications of advanced electronic materials and devices. The series focuses on topics ranging from the theory, modeling, devices, performance and reliability of electron and ion integrated circuit devices and interconnects, insulators, metals, organic materials, micro-plasmas, semiconductors, quantum-effect structures, vacuum devices, and emerging materials. Applications of devices in biomedical electronics, computation, communications, displays, MEMS, imaging, micro-actuators, nanoelectronics, optoelectronics, photovoltaics, power ICs and micro-sensors are also covered.

Books published in the series include research monographs, edited volumes, handbooks and textbooks. The books provide professionals, researchers, educators, and advanced students in the field with an invaluable insight into the latest research and developments.

Topics covered in the series include, but are by no means restricted to the following:

- Integrated circuit devices
- Interconnects
- Insulators
- Organic materials
- Semiconductors
- Quantum-effect structures
- Vacuum devices
- Biomedical electronics
- Displays and imaging
- MEMS
- Sensors and actuators
- Nanoelectronics
- Optoelectronics
- Photovoltaics
- Power ICs

For a list of other books in this series, visit www.riverpublishers.com

Rad-hard Semiconductor Memories

Editors

Cristiano Calligaro

RedCat Devices
Italy

Umberto Gatti

RedCat Devices
Italy

River Publishers

Published, sold and distributed by:
River Publishers
Alsbjergvej 10
9260 Gistrup
Denmark

River Publishers
Lange Geer 44
2611 PW Delft
The Netherlands

Tel.: +45369953197
www.riverpublishers.com

ISBN: 978-87-7022-020-0 (Hardback)
       978-87-7022-019-4 (Ebook)

This book is dedicated to the memory of Prof. Calogero "Rino" Pace,
a smart scientist and good friend.

Contents

Preface

Integrated Circuits (ICs) for space have gained an increasing popularity as far as the satellite market expanded. Such circuits are subjected to a constant bombardment of high-energy particles and in general to harsh operating conditions in terms of radiation, temperature and mechanical stress.

Rad-hard Semiconductor Memories is intended to be a journey in the complex labyrinth of radiation effects in semiconductor materials (silicon dioxide, doped silicon, manufacturing processes) producing a very wide spectrum of failures (from hard errors to soft errors) and during this journey foundations of Radiation Hardening by Design (RHBD) will be provided step-by-step following the canonical floor-plan that microelectronic designers use for realizing ASICs: from the architecture (RHBD at Architectural Level) to circuit block topology (RHBD at Circuit Level) and down to the physical layout design (RHBD at Layout Level).

As the reader will see, designing a rad-hard component is the arena of compromises where performance and area consumption are constantly under threat when trying to make resilient the device.

This kind of design is not a new activity and several publications are already available, mainly on the digital and memory side, but a structured overview is still missing. With this book we would like to make a first attempt to extract an organic presentation from raw material (radiation effects and manufacturing processes) to physical blocks (SRAM arrays, peripheral blocks) and, passing through examples of existing components widely tested under irradiation, provide some guide lines on how a rad-hard memory, whenever it is volatile or non-volatile, must be shaped.

Rad-hard Semiconductor Memories begins with the description of the effect of radiations on semiconductors and the presentation of the design techniques needed to make memories resilient (Chapters 1 and 2). The central part (Chapters 3 to 6) is devoted to the practical implementation of SRAMs, OTPs and Flash memories, which is the outcome of more than a decade of rad-hard component design that we did in our company (RedCat Devices).

Chapters 7 and 8 is a focus on rad-hard resistive memories which represent a valid alternative to conventional memories even if they are still on a development phase. The final chapter is an overlook on emerging technology for non-volatile memories in graphene nanomaterials.

Cristiano Calligaro and Umberto Gatti

Acknowledgements

Part of this book is the result of research activities in the framework of FP7 262890 SkyFlash Project (Chapters 5 and 6) and Horizon2020 640073 R2RAM Project (Chapters 7 and 8). The authors acknowledge the European Commission, the Research Executive Agency (REA) and the former project officers Traian Branza (for SkyFlash) and Giuseppe Daquino (for R2RAM).

The work related to part in Chapter 1 in this book was supported by the European Space Agency (ESA/ESTEC Contract 4000111630/14/NL/PA), and the Academy of Finland under the Finnish Centre of Excellence Programme 2012–2017 (Project No 2513553, Nuclear and Accelerator Based Physics).

List of Contributors

Alexandre Bosser, *Space Technology Group, Department of Electronics and Nanoengineering, Aalto University, Finland*

Anna Arbat Casas, *XGLab S.R.L., Milano, Italy*

Ari Virtanen, *Department of Physics, University of Jyväskylä, FI-40014 Jyväskylä, Finland*

Arto Javanainen, *1. Department of Physics, University of Jyväskylä, FI-40014 Jyväskylä, Finland;*
2. Electrical Engineering and Computer Science Department, Vanderbilt University, Nashville, TN 37235, USA

Christian Wenger, *IHP, Im Technologiepark 25, 15236 Frankfurt/Oder, Germany*

Cristiano Calligaro, *RedCat Devices, Italy*

Evgeny Pikhay, *TowerJazz, Israel*

Luigi Dilillo, *CNRS (Centre National De La Recherche Scientifique)/ University of Montpellier/LIRMM (Laboratoire d'Informatique, de Robotique et de Microélectronique de Montpellier), France*

Paolo Bondavalli, *Thales Research and Technology, France*

Umberto Gatti, *RedCat Devices, Italy*

Yakov Roizin, *TowerJazz, Israel*

List of Figures

List of Tables

List of Abbreviations

ASIC	Application Specific Integrated Circuit
ATD	Address Transition Detection
BBT	Band-to-Band Tunneling
BEOL	Back-End-Of-Line
BIOS	Basic Input-Output System
BL/BLN	Bit line/Negated bit line
BPSG	Boro-Phospho-Silicate-Glass
BRS	Bipolar Resistive Switching
CAS	Column Access Strobe
CEONOS	Cost Efficient Oxide Nitride Oxide Silicon
CF	Conductive Filament
CG	Control gate
CHE	Channel Hot Electrons
CKB	CheckerBoard
CNT	Carbon NanoTubes
CPU	Central Processing Unit
CVD	Chemical Vapor Deposition
DDR-SDRAM	Double Data Rate-Synchronous DRAM
DOS	Density Of States
DRAM	Dynamic Random Access Memory
DPSRAM	Dual Port SRAM
DUT	Device Under Test
ECC	Error Correction Code
EDAC	Error Detection and Correction
EDP	Erase During Programming
EEPROM	Electrically Erasable Programmable Read-Only Memory
EGR	Enhanced Guard Ring
ELT	Edge less transistors
EMI	ElectroMagnetic Interference
EPROM	Erasable Programmable Read-Only Memory
ERD	Emerging Research Devices

FeRAM	Ferroelectric RAM
FEOL	Front-End-Of-Line
FET	Field Effect Transistors
FG	Floating gate
FOX	Field Oxide
FPGA	Field Programmable Gate Array
FSG	Fluorinated Silicon Glass
GFM	Graphene Flash Memory
GO	Graphene Oxide
GOX	Gate Oxides
GST	Germanium-Antimony-Tellurium
HRS	High-Resistive State
I-STI	Inverse STI
IC	Integrated Circuits
ILD	Inter-Level Dielectric
ISPVA	Incremental Step Pulse with Verify Algorithm
ITO	Indium Tin Oxide
ITRS	International Technology Roadmap for Semiconductors
LDD	Lightly Doping Dose
LET	Linear Energy Transfer
LRS	Low-Resistive State
MBU	Multiple Bit Upset
MCU	Micro-Controller Unit
MIM	Metal-Insulator-Metal
MLC	Multi-Level Cell
MLG	Multi-Layered Graphene
MLG-FM	MLG-Flash Memory
MMU	Memory Management Unit
MOSFET	Metal-Oxide-Semiconductor Field Effect Transistor
MRAM	Magnetoresistive RAM
MTP	Multi-Time Programmable
N/P-MOS	N/P-type Metal-Oxide-Semiconductor transistor
NROM	Nitride ROM
NVM	Non-Volatile Memory
OxRRAM	Oxide Resistive RAM
OBC	On-Board Computer
ONO	Oxide-Nitride-oxide
OTP	One Time programmable

PCM	Phase-Change Memory
PROM	Programmable Read Only Memory
QPC	Quantum-Point Contact
R-GO	Reduced Graphene-Oxide
RAM	Random Access Memory
RAS	Row Access Strobe
ReRAM	Resistive RAM
RFID	Radio-Frequency Identification
RHBD	Radiation-Hardened-by-Design
RHBP	Radiation-Hardened-By-Process
RHBS	Radiation-Hardened-By-Shielding
ROM	Read Only Memory
SDRAM	Synchronous DRAM
SEC	Single Error Correction
SECDED	Single Error Correction Double Error Detection
SEE	Single Event Effects
SEFI	Single Event Functional Interrupt
SEGR	Single Event Gate Rupture
SEL	Single Event Latch-up
SEM	Scanning Electron Microscope
SET	Single Event Transitions
SEU	Single Event Upset
SILC	Stress Induced Leakage Currents
SLC	Single-Level Cell
SNOS	Silicon-Nitride-Oxide-Silicon
SoC	System-on-Chip
SONOS	Silicon-Oxide-Nitride-Oxide-Silicon
SPSRAM	Single Port SRAM
SRAM	Static Random Access Memory
STI	Shallow Trench Isolation
STTRAM	Spin-Transfer Torque RAM
TAS MRAM	Thermally Assisted Magnetic RAM
TEM	Transmission Electron Microscopy
TEOS	Tetra-Ethyl-Ortho-Silicate
TID	Total Ionizing Dose
UV	Ultraviolet
USG	Undoped Silicon Glass
XPS	X-ray Photoelectron Spectroscopy
WL	Word line

1

Space Radiation Effects in Electronics

Luigi Dilillo[1], Alexandre Bosser[2], Arto Javanainen[3,4] and Ari Virtanen[3]

[1]CNRS (Centre National De La Recherche Scientifique)/
University of Montpellier/LIRMM (Laboratoire d'Informatique,
de Robotique et de Microélectronique de Montpellier), France
[2]Space Technology Group, Department of Electronics and Nanoengineering,
Aalto University, Finland
[3]Department of Physics, University of Jyväskylä, FI-40014 Jyväskylä,
Finland
[4]Electrical Engineering and Computer Science Department,
Vanderbilt University, Nashville, TN 37235, USA

> "I am accepted in both sietch and village, young Master. But I am in
> His Majesty's service, the Imperial Planetologist."
>
> Frank Herbert, Dune

Abstract

This chapter introduces the topic of the effects of harsh radiation environments, first on electronic devices in general, and next on memory components, the case study of the book. Ionizing radiation is ubiquitous in space, in the form of charged particles, coming mainly from the Sun, supernovae explosions and other galaxies, and which can be trapped and concentrated by planetary magnetic fields. When striking electronic components, these energetic particles cause both cumulative and single-event (spontaneous) effects. The former induce a degradation in performance, such as increased leakage currents, which can eventually lead to the complete loss of device functionality. The latter are generally non-destructive, and in memory components, may induce bit flips. Electronic devices used in harsh radiation environments are almost systematically characterized with accelerated tests in irradiation facilities, which use radiation beams to reproduce the target environment. For space applications, single-event characterization is typically done using particle accelerators which provide heavy ions and/or protons.

1.1 Space Radiation Environment

In the vacuum of space, energetic particles can travel over great distances without energy loss. In outer space, away from any radiation sources (i.e. stars) and intense magnetic dipoles, the amount of particle radiation is very small. However, in proximity of the Sun and the Earth, the particle radiation flux is significant. In this chapter, the main focus will be the radiation environment from the Sun to the upper atmosphere of the Earth. In this region of space, particle radiation can be divided into three different categories: (1) solar energetic particles, directly originating from the Sun, (2) particles trapped in the Earth's magnetic field, and (3) Galactic Cosmic Rays (GCR). These categories are discussed in the following chapter, along with the monitoring and forecasting of space weather, and its implications for our everyday life.

1.1.1 The Sun

Near the Earth, the Sun is the dominant radiation source; the majority of particle radiation present in the interplanetary space near the Earth or trapped in the Earth's magnetosphere originates from the Sun. In addition to electromagnetic radiation, the solar wind mostly consists of energetic electrons and protons. In the solar wind, the plasma travels through space with an average velocity of approximately 400 km/s, and the average proton densities, in the absence of any major solar events, are in the order of 1^{-10} protons/cm^3 [1]. This yields average proton fluxes of $4 - 40 \cdot 10^7 \text{cm}^{-2}\text{s}^{-1}$.

1.1.2 The Sunspot Cycle

The Sun's activity fluctuates with a period of approximately 11 years, which is called the solar cycle. The activity of the Sun is closely related to the number of sunspots observed in the Sun's surface. The first European observations of the sunspots were made in the 17th century, but the daily records were started in 1749 [2]. There are some scarce records available, dating back to the year 1610. The counting of these cycles was started in the 1760s, when the occurrence of the first official full cycle has been agreed; in 2017, the 24th cycle is ongoing. The monthly average of the sunspot number from Ref. [2]

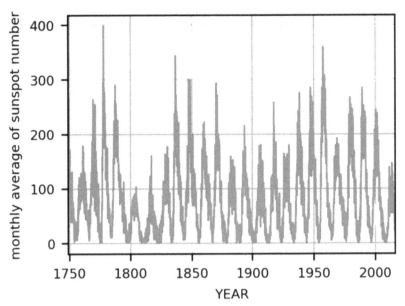

Figure 1.1 Monthly average of sunspot number from the beginning of 1749 to September 2016 as taken from Ref. [2].

is presented in Figure 1.1. The "11-year" periodicity in the sunspot number is clearly seen in the figure.

1.1.3 Solar Flares and Coronal Mass Ejections

Apart from the fairly predictable solar cycle discussed above, there are also unpredictable phenomena occurring in the Sun. Sometimes, interactions between the hot plasma and the magnetic fields on the surface of the Sun lead to the expulsion of vast quantities of plasma. These eruptions are called solar flares, and the most massive solar flares are called Coronal Mass Ejections (CMEs). CMEs are orders of magnitude larger compared to regular solar flares. In CME events, a large part of the Sun's corona is ejected from the Sun with velocities up to thousands of kilometres per second, with ejecta mass up to 10^{13} kg [5].

The solar flares mainly consist of electrons and protons, but there are also heavier ions, such as oxygen and iron, present in these shockwaves. In Figure 1.2, the average particle fluxes in the solar wind for oxygen and iron ions are presented as they have been measured by the Solar Isotope Spectrometer [6] on board the Advanced Composition Explorer spacecraft [3].

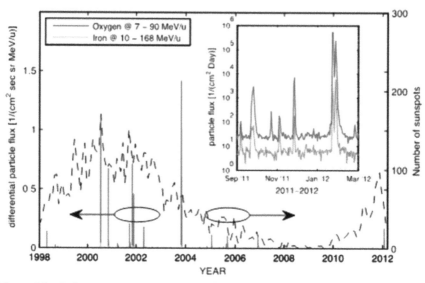

Figure 1.2 Daily average of solar energetic particle fluxes from 1998 to March 2012 for oxygen and iron ions measured by the Solar Isotope Spectrometer on the Advanced Composition Explorer spacecraft [3], with monthly average sunspot number taken from Ref. [2]. Particle flux data are taken from Ref. [1]. Figure taken from [4].

These data are available online from Ref. [1]. The graph presents the data taken between two solar maxima around years 2001 and 2012. In the graph, the solar activity is presented by the monthly averaged sunspot number from Ref. [2]. The data clearly demonstrate the increased occurrence of bursts in the heavy-ion fluxes during the high activity season of the Sun. The variability in the ion fluxes is more easily observed in the inset graph. The observed peaks are associated with solar flares, and the highest ones are due to CMEs. Although the amount of heavier ions in CMEs is much lower than that of the solar wind protons or electrons, they can cause problems in the electronics more easily due to their higher ability to ionize matter (i.e. stopping force). More discussion on the particle–matter interactions in materials and particle-induced effects in electronics is presented below.

1.1.4 Trapped Particles – Van Allen Belts

The Earth's magnetic field and atmosphere prevent the energetic charged particles from directly reaching the Earth's surface. As the fast-moving charged particles enter the Earth's magnetosphere, they experience tangent Lorentz forces which bend their trajectory. This trajectory bending can lead to particle trapping in the magnetosphere, thus creating regions with high radiation fields known as the Van Allen belts [7]. These belts mostly consist of energetic protons and electrons. There are two distinct regions called the inner and outer belts: the inner belt is generally referred to as the proton belt and the outer one as the electron belt, because these regions are dominantly populated by these particles.

1.1.5 South Atlantic Anomaly

The Earth's magnetic poles (North pole: 85.01°N, 132.66°W, South pole: 64.43°S, 137.32°E [8]) diverge from the geographical poles. In addition, the magnetic poles are not antipodal, in contrast to the geographical poles which are antipodal by definition [8]. This misalignment causes eccentricity in the Van Allen belts with respect to the Earth's surface. Hence, there is an increased amount of energetic particles reaching very low altitudes in the South Atlantic region east of Brazil. This region is called the South Atlantic Anomaly (SAA). The SAA is illustrated in Figure 1.3, where the fluxes

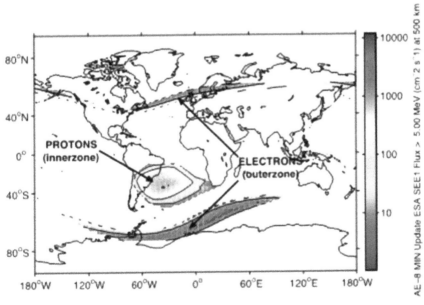

Figure 1.3 Proton (100 MeV) and electron (5 MeV) intensities at 500 km altitude estimated by the AP8-MIN and AE8-MIN models, respectively [19, 20]. The South Atlantic Anomaly can be clearly seen in the south–east coast of Brazil, where there is a distinct region of energetic protons reaching altitudes of 500 km. Data taken from the ESA's Space Environment Information System (SPENVIS) in Ref.[9]. Figure taken from Ref. [4].

of >100 MeV protons and >5 MeV electrons at an altitude of 500 km are presented as taken from Ref. [9]. One should note that in this figure, the SAA region consists only of protons at the given energies. The SAA exposes high-altitude aircrafts to higher radiation levels than other regions around the globe. Moreover, the SAA poses a severe threat to satellites orbiting the Earth.

1.1.6 Galactic Cosmic Rays

Satellites orbiting the Earth below the Van Allen belts, in Low Earth Orbit (LEO), are experiencing relatively low and stable radiation fluxes, with occasional disruptions by high solar activity (i.e. solar flares and CMEs, see discussion above) and passages through the South Atlantic Anomaly. The radiation environment within and outside the radiation belts is much more irregular and influenced by the solar activity.

Aside from the solar contribution to the radiation environment, a relatively steady flux of ions at very high energies can be identified. These particles are called Galactic Cosmic Rays (GCR), as they originate from supernovae in other galaxies. Their energies can reach up to several hundreds of GeV/u. By using the CREME code, available online at Ref. [10], the radiation environment due to GCRs can be estimated. GCR spectra at the Geostationary Earth Orbit (GEO) or in near-Earth interplanetary space are presented in Figure 1.4.

In Figure 1.4, the differential fluxes of selected GCR particles are given as a function of energy per nucleon. The fluxes are defined after 2.54 mm of aluminium, which are a typical shielding thickness and material for spacecraft. The selected particles are proton, helium, carbon, silicon and nickel. The particle fluxes are plotted according to the estimations for solar minimum and maximum. The difference in the GCR spectra, presented in Figure 1.4, between the solar maximum and minimum, is due to the variation of the intensity of the interplanetary magnetic field, which is correlated to the solar activity. High solar activity during the solar maximum gives rise to higher solar magnetic fields, and thus lower-energy GCRs are deflected away from the inner planets of the Solar System.

Among GCRs, protons are the most abundant species, as can be seen in Figure 1.4. This trend is also illustrated in Figure 1.5, where the relative abundances of particle species are presented as a function of atomic number. This graph shows the rapid drop in particles with atomic numbers above that of iron ($Z \sim 26$). This is due to the maximum in the nuclear binding energy for atomic masses around $A = 60$. For simulation purposes, this results in a practical maximum value for the electronic stopping force in silicon of GCRs

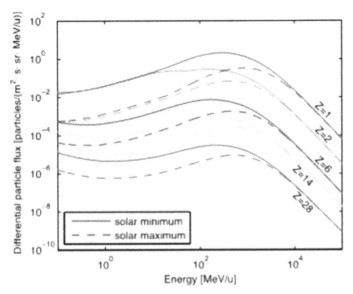

Figure 1.4 Differential flux of selected galactic cosmic ray particles after 2.54 mm of aluminium as a function of energy per nucleon in GEO or near Earth interplanetary space during the solar minimum (solid line) and the solar maximum (dashed line). Data are taken from [10]. Figure taken from Ref. [4].

to be approximately 30 MeV/(mg/cm^2). More discussion of the electronic stopping and the radiation effects and their testing in electronics are presented below.

1.1.7 Space Weather

The space radiation environment (also called "space weather"), is constantly monitored by various dedicated satellite systems, such as the Geostationary Operational Environmental Satellites (GOES) and the Solar and Heliospheric Observatory (SOHO). For instance, the Space Weather Prediction Center (SWPC) of the National Oceanic and Atmospheric Administration (NOAA) provides a service for real-time monitoring and forecasting of the space environment. Their webpages can be found at http://www.swpc.noaa.gov/. Real-time information about the space radiation environment is crucial for the safety of modern satellite systems, and also to some extent for electronic and electrical systems on the ground. While smaller solar flares are only a threat to spacecraft, the most intense flares (CMEs) can create geomagnetically induced currents (GIC) on the Earth's surface which can also affect the power

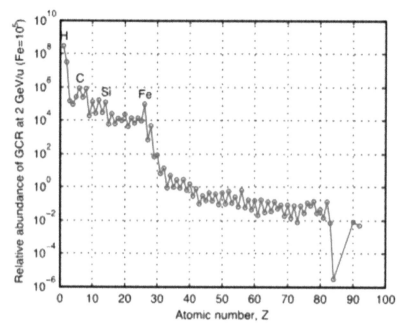

Figure 1.5   Relative abundance of different ion species in galactic cosmic rays as a function of atomic number at energies ∼2 GeV/u. Data are taken from [10]. Figure taken from Ref. [4].

grids, telecommunication networks, pipelines and railways. The GICs and their effects in the power lines in Finland have been discussed in Ref. [11]. The solar flares also increase the radiation levels in the atmosphere due to the increase in trapped particles in the Van Allen belts. The most well-known and easily observable indication of increased atmospheric radiation are the northern and southern lights, which are also known as the Aurorae Borealis and Aurorae Australis. These phenomena are caused by interactions between space radiation and atmospheric gas at high altitude.

1.1.8 Atmospheric and Ground-Level Radiation Environments

Particle radiation is not restricted only to space. Natural and artificial sources of particle radiation may be found on Earth at high altitude, on the ground and underground, and have been shown to affect electronics. This section gives a brief introduction to the radiation sources existing at the ground level and in the Earth's atmosphere. First, we present a discussion on the natural sources, followed by a short contemplation of the artificial sources.

1.1.9 Cosmic Rays

Although the Earth's magnetic field deflects the majority of the particle radiation coming from space, there is still a continuous shower of energetic particles present at high altitudes and even at the ground level due to the GCRs and the solar activity. The atmospheric radiation was shown to increase with altitude already 100 years ago by Victor Hess, who also proved the radiation to be mostly of cosmic origin [12].

The probabilities of the primary cosmic ray particles (protons and heavy ions) themselves penetrating the atmosphere are extremely low, but as they interact with atoms in the upper atmosphere, secondary energetic particles are released, such as neutrons and muons. These secondaries interact relatively weakly with matter, and they are able to reach ground level, and to some extent even penetrate the Earth. Neutrons, being neutral particles, do not interact with matter via Coulomb forces and thus no primary ionization is produced by them. In turn, neutrons can indirectly produce ionizing effects via scattering with nuclei, elastically or inelastically.

Early in 1979, Ziegler and Lanford in Ref. [13] suggested that the anomalous soft errors observed in random-access-memories (RAM) were caused by atmospheric neutrons. They also predicted this effect, at that time considered quite minute, to become more pronounced in the future along with the technological evolution. In his review in Ref. [14], Normand discussed the impact of the two papers, Refs. [13], [15], on the activities of the electronics industry. Normand criticized that the susceptibility of electronics to errors induced by atmospheric neutrons remained publicly unrecognized. Even though the vendors (e.g. IBM [16]) made extensive studies about the subject early on, they published the results much later. Normand's paper presents several examples of soft error observations in memories and explains them with the atmospheric neutrons. The impact of atmospheric neutrons on scaled technologies has been discussed recently by Ibe et al. in Ref. [17], where, based on their simulations, they confirm Ziegler's postulate that the effect becomes more pronounced along with the technology scaling. Some anecdotal suggestions have been made by representatives from electronics industry about the increasing possibility of cosmic rays causing failures in the evolving automotive electronics if not taken into account in the design [18]. This aspect is also discussed in Ref. [19 and references therein].

Atmospheric radiation is more of a problem in avionics. After the solar flare occurred in late January 2012 (see the inset of Figure 1.2), several airlines were forced to divert their flights from the polar routes due to increased radiation in the atmosphere [20]. The effects of cosmic rays on avionics has been discussed, e.g. in Refs. [21, 22].

Apart from neutrinos, muons are the most abundant of the cosmic ray secondaries present on the ground [23]. Muons are a charged particle with the same charge as electron, and mass corresponding approximately 200 times the electron mass. The first results on muon-induced errors in microelectronics were published by Sierawski et al. [24], where an artificially produced muon beam was used. Additional results from irradiations of SRAM memory components with low-energy muon beams were reported in [25, 26]. These results suggest that while low-energy muons are capable of inducing single-event upsets in 65 nm and smaller technology nodes in a lab environment, atmospheric muons (which have relatively high energy and low electronic stopping force) are incapable of doing so. However, these results predict the susceptibility of components to become higher in the future technologies.

1.1.10 Radionuclides in the Soil

Although silicon is the second most abundant element (by weight) in the Earth's crust [27], the use of this excellent metalloid in semiconductor industry has exhibited some unexpected problems in the past. Trace radioactive elements are present everywhere in the soil; thus, all bulk materials contain small traces of radioactive particles, even after refinement. The most concern is due to U-238 and Th-232, and their radioactive daughter nuclides, such as Rn-220 and Rn-222.

In 1979, May and Woods reported in Ref. [15] that alpha particles emitted from the packaging materials were causing soft errors in random access memories (RAM) and charge-coupled-devices (CCD). It was shown that the packages used at that time contained these radioactive elements in the order of some parts per million. In 1982, the metal layers within integrated circuits was demonstrated to be another possible source of radioactivity [28]. After these revelations, efforts were made to eliminate these sources of radiation and mitigate their effects, typically by enhancing the purity of the materials and/or implementing error detection and correction (EDAC) in these devices. In addition, protective coatings on the device surface have been introduced in order to minimize the problem [29]. However, these "auxiliary" materials are not the only source of radiation. Even after silicon goes through a thorough refining process, trace amounts (∼parts per billion) of these unwanted radioactive elements remain present, which in turn have been demonstrated to cause soft errors in modern memories [30, and references therein].

Radon is a radioactive gas which is naturally found around certain types of soil, and is a well-known health hazard. A recent study presented in Ref. [31] has shown that ambient radon is very unlikely to induce errors in microelectronics via diffusion through the packaging. Nevertheless, while being an inert atom, Rn can diffuse into materials. Also, its daughter nuclides, usually positively charged ions, can adhere to surfaces [32]. These radioactive elements have to be taken into account in devices with bare dies such as particle detectors.

These problems, caused by the radioactive residues, are nearly impossible to avoid completely. Only very strict control of the materials used in manufacturing and robust EDAC techniques are effective in assuring the tolerance to these radiation sources.

1.1.11 Thermal Neutrons

Eventually, if they are not captured by target material, the secondary neutrons (produced by either the aforementioned cosmic rays, or decay of radionuclides in the soil) lose most of their energy and become thermalized. These so-called thermal neutrons have an average kinetic energy of $\sim$25 meV, which is the most probable energy for a free particle at room temperature ($\sim$290 K). It was discovered by Fermi et al. (see Ref. [33, and references therein]) that slow neutrons are more effectively interacting with matter than high-energy ones. Thermal neutrons can initiate nuclear reactions, if they are captured by target nuclei. Depending on the target atom, the reaction products can be either gamma rays or particle radiation. If particle radiation is released in the reaction (especially for fission products), a large amount of ionization will be generated around the site of the interaction, which can be a source of errors in electronic components. The most notorious example of this phenomenon is the thermal neutron-induced fission of boron-10 described by

$$^{10}_{5}B + {}^{1}_{0}n \rightarrow {}^{7}_{3}Li + {}^{4}_{2}He + \gamma$$

This is due to $^{10}_{5}B$ having an exceptionally high cross-section for thermal neutron capture, when compared with other materials. This has been shown, in Ref. [34], to be a considerable source of errors in Static Random Access Memories (SRAM). Due to the relatively high fluxes of low-energy neutrons and their high penetration capability, the only way to mitigate this problem is to avoid using $^{10}_{5}B$ in the devices. Boron is a very widely used dopant in semiconductors, and is present in the borophosphosilicate glass (BPSG) commonly used as an insulator in standard manufacturing processes. Although naturally $^{11}_{5}B$ is the most abundant ($\sim$80%) boron isotope, the boron used in manufacturing has to be carefully refined to eliminate the $^{10}_{5}B$, in order to minimize the problems caused by thermal neutrons.

1.1.12 Artificial Radiation Sources

In addition to natural sources, there are a lot of man-made radiation sources at ground level; radiation from these sources can be a reliability issue for electronic systems. A selected list of man-made radiation sources is given below.

- accelerators, e.g. cyclotrons (heavy ions, protons, neutrons, electrons), synchrotrons (X-rays) and X-ray tubes;
- lasers;
- radionuclides (photons, electrons, neutrons, protons, heavy ions);
- nuclear power plant (photons, neutrons);
- nuclear weapons (e.g. Operation Dominic [35, 36]).

Extra care has to be taken of electronic systems used in the vicinity of these artificial radiation sources, in order to assure their operation. For example, in nuclear power plants or accelerators, such as the Large Hadron Collider at CERN, electronics failures must be minimized to ensure the reliable and safe operation of the whole facility. The radiation sensitivity of exposed electronics must be tested, and in some situations, equipment must be implemented using radiation-hard parts (RadHard). The use of accelerators, namely cyclotrons, and enriched radionuclides, in the radiation effects testing of electronics is discussed later in this book.

1.2 Radiation Effect in Materials and Devices

1.2.1 Energetic Charged Particles and Matter

While an energetic charged particle traverses matter, it loses energy via different mechanisms. A simple illustration of energy loss is given in Figure 1.6, where an ion with initial energy of E penetrates a slab of material with a thickness of $\triangle x$. The final energy of the ion is $E - \triangle E$. If the thickness is considered as infinitesimal, dx, the energy loss $\triangle E \rightarrow dE$. From this, the total stopping force is defined as

$$-\left.\frac{dE}{dx}\right|_{total} = \sum_{i} -\frac{dE}{dx},$$

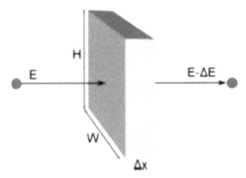

Figure 1.6 Illustration of the energy loss of energetic particle with initial energy of E after passing through a slab of material with thickness of $\triangle x$.

where i denotes the different energy loss mechanisms according to [37], which are listed below:

1. Excitation and ionization of target atoms;
2. Projectile excitation and ionization;
3. Electron capture;
4. Recoil loss ("nuclear stopping") and
5. Electromagnetic radiation.

Additionally, part of the energy loss can also go to the following reactions:

6. Nuclear reactions and
7. Chemical reactions

As in this work, primarily the interactions between heavy ions and matter are considered at energies of hundreds of keV/u to tens of MeV/u, mechanisms 1–4 are the most relevant ones. Moreover, this chapter concentrates on mechanisms 1 and 2 as they constitute most of the electronic energy loss, which dominate above all others in the energy ranges and ion species studied in this work. Nuclear reactions (6) are discussed, as there is evidence of their contribution in the errors observed in modern electronics. Nuclear stopping (4) is also briefly discussed.

Particle radiation can change the physical properties of the target material, both temporarily and permanently. Usually, ionizing radiation (processes 1–3) is considered to be non-destructive, as it is mainly breaking the covalent bonds in the target material. Depending on the material, these broken bonds may be self-reassembled immediately after their creation, or can be repaired by high-temperature annealing. In reality, heavy ions also create permanent changes in the target. This is, on the one hand, due to the high density of ionization caused by the high-Z ions, which induces material modification. On the other hand, energetic target recoils, due to nuclear stopping (4), may modify the atomic structure of the target. In some cases, depending on their nature, these atomic defects can also be repaired by high-temperature annealing. The effects of energetic heavy ions in materials and electronic devices are discussed below.

1.2.2 Stopping Nomenclature

Before proceeding, the terminology of the energy loss will be clarified. Typically, the energy loss of an ion in matter is called the stopping power. In the past, there has been intensive discussions on the terminology, whether the term "stopping power" should be replaced by stopping force [38]. In this text, the term "stopping force", or simply "stopping", is used.

1.2.3 General Theory for Electronic Stopping

When considering the mean energy loss per unit length for a projectile due to the collisions with target electrons (i.e. electronic stopping force), it is generally written as:

$$-\frac{dE}{dx}\bigg|_{elec} = NZ_2 S = \frac{1}{4\pi\epsilon_0^2}\frac{Z_1^2 q^4}{m_e v^2} NZ_2 L = 3.0705 \cdot 10^{-4}\frac{Z_1^2 Z_2}{A_2 \beta^2}L, \quad (1.1)$$

where

$Z_{1,2}$ are the atomic numbers for the projectile (1) and target (2);

A_2 is the mass number of the target atom;

N is the atomic density of the target material;

v is the velocity of the projectile;

S is the electronic stopping cross section in units of $[energy \times area]$;

q and m_e are the elementary charge and the electron rest mass, respectively;

ϵ_0 is the vacuum permittivity and

L is the theory dependent dimensionless stopping number (see below).

Furthermore, $\beta = v/c$ and c is the speed of light in vacuum. The electronic stopping force in Equation (1) is given in units of $MeV/(mg/cm^2)$.

The sophisticated reader might notice that this is the unit for mass stopping force and not for stopping force (MeV/μm). Let us clarify here that in this work, the concepts of mass stopping force and stopping force are used interchangeably, although technically there is a difference. Strictly speaking, the mass stopping force is the stopping force divided by the density of the target material (i.e. $-\frac{1}{\rho}\frac{dE}{dl}$).

In the following sections, the fundamental stopping theories by Bohr, Bethe and Bloch are introduced. Also, description of a semi-empirical model developed in this work is given below.

1.2.4 Stopping Theories and Semi-Empirical Models

Probably, the most widely known and used notation for the electronic stopping force is the Bethe formula, where the stopping number is

$$L_{Bethe} = \ln \frac{2m_e v^2}{\hbar\omega_0} \tag{1.2}$$

This equation is based on the quantum mechanical treatment of the interaction between the projectile and the target electrons. It should be noted that this equation is valid when $2m_e v^2 \gg \hbar\omega_0$. A comprehensive derivation of Equation (2) is presented in Refs. [37, 39].

Another well-known notation for electronic stopping force is from the Bohr's classical theory. Niels Bohr formulated his theory on the decrease of velocity of moving electrified particles on passing through matter in 1913 [40], which he later revised in Refs. [41, 42]. The Bohr's model is approximately valid for velocities

$$\beta < 2Z_1\alpha, \tag{1.3}$$

where $\alpha \sim \frac{1}{137}$ is the fine structure constant. The lower limit of validity is set by $\beta \gg \alpha$. In the classical picture, the target electrons are treated as harmonic oscillators with resonance frequencies of ω_i. The collisions are considered as elastic. From this treatment on the interaction between the impinging energetic ion and an individual target electron, Bohr's stopping number becomes

$$L_{Bohr}(\xi_i) = \ln(C\xi_i), \tag{1.4}$$

where $\xi_i = \dfrac{m_e v^3}{Z_1 \cdot \alpha \cdot c \cdot \hbar \cdot \omega_i}$ and $C = 2e^{-\gamma} \cong 1.1229$. Here, γ is the Euler-Mascheroni constant usually known as the Euler's constant (not to be confused with the Euler's number $e = 2.71828...$). By summing over all electrons in the target atom, we get

$$L_{Bohr} = \sum_i f_i \ln(C\xi_i) = \ln(C\xi), \qquad (1.5)$$

where f_i is the weighting factor corresponding to the individual electron ($\sum_{i=1}^{Z_2} f_i = 1$). Now, we can write

$$\xi = \frac{m_e v^3}{Z_1 I \alpha c}, \qquad (1.6)$$

where $I = \hbar \omega_0$ is the material-dependent mean excitation energy corresponding to the mean frequency.

At low projective velocities, there is a drawback in Equation (5), i.e. $\xi \leq \frac{1}{C} \Rightarrow L_{Bohr} \leq 0$. Solution to this problem has been presented in Ref. [37], where the works of Lindhard and Sorensen [43] are articulated. In this derivation, the Bohr's stopping number becomes

$$L'_{Bohr} = \frac{1}{2} \ln[1 + (C\xi)^2]. \qquad (1.7)$$

There is a myriad of semi-empirical approaches for estimating the electronic stopping force.

1.2.5 Nuclear Stopping Force

The nuclear stopping is the part of the projectile's energy loss due to Coulombic interaction between the target and the projectile nuclei. With a similar treatment as for the electronic stopping above, a general equation for nuclear stopping force can be written as

$$-\left.\frac{dE}{dx}\right|_{nucl} = \frac{1}{4\pi\epsilon_0^2} \frac{Z_1^2 Z_2^2 q^4}{M_2 v^2} N L_{nucl}, \qquad (1.8)$$

where the notations are the same as in Equation (1) and, in addition, M_2 is the mass of the target nucleus. The nuclear stopping relative to electronic stopping can be estimated by dividing Equation (8) with Equation (1) yielding a ratio of

$$\frac{-\frac{dE}{dx}\big|_{nucl}}{-\frac{dE}{dx}\big|_{elec}} = \frac{m_e}{M_2} Z_2 \frac{L_{nucl}}{L_{elec}} \approx 2.7 \cdot 10^{-4} \frac{L_{nucl}}{L_{elec}}. \qquad (1.9)$$

From this, it can be seen that the contribution of the nuclear stopping in the total stopping force is several orders of magnitude lower than that of the electronic stopping. This holds at projectile velocities typically used in radiation effects testing discussed later in the text. The ratio of the stopping numbers, and thus the nuclear stopping, becomes significant only at low ion velocities [37]. Even though in the nuclear stopping process the energy lost by the projectile is transferred mainly to the kinetic energy of the target nucleus, and the recoil loses its energy partly via electronic stopping, the ionization effect is typically considered to be less significant. The effect of nuclear stopping is typically attributed to the displacement damage (DD) effect, discussed later in the text.

1.2.6 Ion-induced Nuclear Reactions

As mentioned above, ions can lose their energy also by inducing nuclear reactions. The cross sections for the nuclear reactions are much lower than the ones for the other mechanisms discussed above, and they are typically neglected. This is especially the case for heavy ions. Nevertheless, in microelectronics, the products from the nuclear reactions may induce failures. Thus, the ion-induced nuclear reactions are briefly introduced here.

When considering positively charged projectile (Z_1, A_1, R_1) and target (Z_2, A_2, R_2) nuclei, there is a repulsive electromagnetic force acting on them. Here, Z_i, A_i and R_i correspond to the atomic and mass numbers and the radii of the nuclei, respectively. The radius of a nucleus is typically estimated

$$R_i = R_0 \cdot A_i^{1/3} fm, \tag{1.10}$$

where R_0 is an empirical constant. Typically, $R_0 = 1.2 fm$ is used. This constant can vary in the range of $1 \ fm < R_0 < 4.5 \ fm$, depending on the nucleus [44].

In order for a nuclear reaction to occur, the nuclei need to be in close contact. This means that the impact parameter has to be less than the sum of the radii of the nuclei, $R = R_1 + R_2$, and the total kinetic energy in the system has to exceed the potential energy formed by the repulsive Coulombic force. This limit in the potential energy is called the Coulomb barrier, which can be estimated by

$$U_{CB} = \frac{1}{4\pi\epsilon_0} \frac{Z_1 Z_2 e^2}{R} \approx 1.2 \frac{Z_1 Z_2}{A_1^{\frac{1}{3}} + A_2^{\frac{1}{3}}} \ MeV. \tag{1.11}$$

From the conservation laws of momentum and energy, the threshold kinetic energy in the laboratory frame for the projectile to overcome the Coulomb barrier can be written as [44]

$$K_{cbt} = U_{CB} \cdot \left(1 + \frac{A_1}{A_2}\right).$$ (1.12)

For light projectiles, where $A_1 \ll A_2$, the kinetic energy threshold is essentially the same as the Coulomb barrier given by Equation (11). Classical estimation for the maximum total cross section for nuclear reactions is given by

$$\sigma = \pi R^2 \cdot \left(1 - \frac{K_{cbt}}{E_k}\right),$$ (1.13)

where E_k is the initial kinetic energy of the projectile [45].

1.3 Radiation Effects in Semiconductors

This chapter presents different phenomena occurring in materials due to the energy deposition by these particles. From electronics' point of view, the most relevant materials are silicon (Si) and silicon dioxide (SiO_2). In some examples in the text, also other materials may be mentioned.

Typically, in electronics, the generation of electron–hole pairs (e–h) due to the energy deposition of any quantum of ionizing radiation (photon or particle) is considered to be the dominant effect. The atomic displacements are considered to play a minor, but still in some cases, a relevant role. This will be discussed below.

When considering the influence of particle radiation in electronics, typically only the electronic stopping force is considered. The electronic stopping force can be related directly to ionization, whereas nuclear stopping and other mechanisms are only causing ionization indirectly. Nuclear reaction products and elastic recoils can also cause ionization, but usually in case of heavy ions, their contributions are neglected. However, recently, there has been an increasing interest toward their contribution, as discussed in Ref. [14, and references within]. The effects of nuclear reactions in materials and devices are briefly discussed below.

1.3.1 Generation of Electron–Hole Pairs

At room temperatures, the energy gap in the silicon band structure is 1.1 eV [47] and for SiO_2, the corresponding value is 9 eV [48]. These are the minimum energies at which the upmost electrons in the valence band (VB) are excited to the conduction band (CB). The average energies required to generate an e–h pair in Si and SiO_2 are 3.6 eV [49] and 17 ± 1 eV [50], respectively. This is called the mean e–h pair creation energy. When considering electronics and the e–h pairs, induced by radiation, there are three main criteria that need to be fulfilled in order to cause major ionizing radiation effects. These criteria are as follows:

The deposited energy in a single encounter between atomic electron and radiation quantum (i.e. ion, delta-electron or photon) has to exceed the band gap energy to produce an e–h pair. In practice, the number of generated e–h pairs is estimated by dividing the mean energy deposition with the mean e–h pair creation energy.

The density of induced e–h pairs has to exceed the intrinsic free electron density of the target material. Hence, in metals, the effect of ionization is negligible due to the intrinsically high count of free electrons.

An electric field is required in order to collect the excess charge carriers.

1.3.2 Nuclear Reactions

Even if the impinging ion itself is incapable of generating enough e–h pairs via electronic stopping mechanisms to cause disturbance in electronics, it still may induce nuclear reactions as discussed above. In turn, the reaction products may be capable of producing high-density plasma of e–h pairs. This is typically encountered with high-energy protons that exhibit low electronic stopping, but still are capable of inducing errors in the devices. These errors originate from the natSi(p,X)Y nuclear reactions, already mentioned above. An example of this phenomenon is presented in Figure 1.7, where a Single Event Upset (SEU) cross-section curve from Ref. [51] for a 4 Mbit SRAM is plotted as a function of proton energy. In this figure, it is seen that low-energy protons (< 20 MeV) are less capable of producing errors in the device than high-energy protons, for which the cross section (i.e. probability) is saturated. The low-energy behaviour is due to the Coulomb barrier, discussed above, which can be estimated to be 4.3 MeV for protons in silicon. This is roughly observed in Figure 1.7, although the data are not precise enough to make solid conclusions. The saturation in the cross section originates from the limitation in the electronic stopping force for the reaction products. For example, if we consider the above-mentioned reaction, the reaction product that has the highest capability for energy deposition is phosphorus ($Z_1 = 15$). The maximum energy loss value, according to SRIM, for this ion is 15 MeV/(mg/cm^2) that

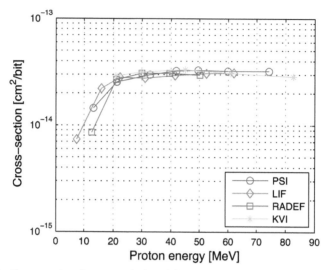

Figure 1.7   Cross section for proton-induced SEUs as a function of proton energy measured from 4 Mbit Atmel AT60142F SRAM used in the SEU monitor [52]. Data taken from Ref. [51].

sets the upper limit for the deposited energy from the reaction. In reality, the probability for natSi(p,γ) P-reaction can be assumed to be lower than that for other reaction channels, but it enables the highest energy deposition.

Another demonstration of the contribution of the nuclear reactions in the energy deposition of protons in silicon is presented in Figure 1.8. In this figure, the energy deposition of 30 MeV protons in silicon target is presented as simulated with the MRED code [10]. The target was a slab of silicon with arbitrarily chosen dimensions of $100 \times 100 \times 20$ μm ($W \times H \times \triangle x$, see Figure 1.6). The MRED code enables the omission of the physics definitions for the nuclear reactions from the simulations. Thus, the effect of nuclear reactions is possible to indicate. The spectrum for energy deposition in Figure 1.8, with only the Coulombic scattering taken into account, exhibits a peak near 60 keV (i.e. 3 keV/μm). This is comparable to the SRIM estimation for the electronic stopping force of silicon for 30 MeV protons, that is, ~3.4 keV/μm. The spread in the spectrum is due to the straggling. The simulation, where also the nuclear reactions are included in the simulations, results in a spectrum with the same primary ionization peak, but in addition, there is a tail reaching up to 10 MeV (0.5 MeV/μm) of energy deposition within the volume. Although the probability of these events is orders of

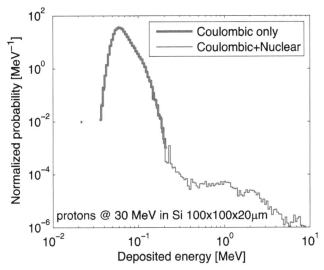

Figure 1.8 MRED simulation of energy deposition by 30 MeV protons in silicon target with size of $100 \times 100 \times 20 \, \mu m$ ($W \times H \times \triangle x$).

magnitude lower than of the primary ionization, in some cases, they may turn out to be detrimental.

1.3.3 Linear Energy Transfer vs. Electronic Stopping Force

When discussing the ion–matter interactions, historically in medical applications and also in radiation testing of microelectronics, the term Linear Energy Transfer (LET) has been used for the average deposited energy by the impinging particle per unit length. Typically, the concepts of LET and electronic stopping force have been used interchangeably. The units of these two are the same, typically given in $MeV/(mg/cm^2)$. However, there is a slight difference between these two, as the electronic stopping force quantifies the mean energy lost by the particle per unit length (in collisions with the target electrons). In case of bulky objects, e.g. old technology electronic components with large sensitive volumes (SV), where the energy is deposited, this is a valid choice. Historically, in radiology, the irradiated tumours have been considered to be bulky, thus justifying the use of mean energy loss as a metric for the mean energy deposition. For a long time, in both communities, considering either radiation effects in the electronics [53–56] or the radiology [57–60], there has been discussions about the validity of LET as a metric and the comparability of LET and electronic stopping force. Although in biology

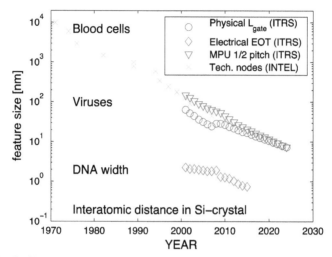

Figure 1.9   Scaling trends of some feature sizes in microelectronics, including physical gate length (Lgate), electrically equivalent oxide thickness (EOT) and half-pitch in MPU presented by the ITRS [62] and development of Intel's® technology nodes [63]. Also approximates of biological and physical feature sizes are illustrated. Figure taken from Ref. [4].

the radiation effects are shown [61] to be rather complicated, the similarity in these two different fields is in the feature sizes, where the radiation effects take place. In radiology, the noteworthy effects are considered to occur in very small dimensions, e.g. width of a DNA double strain, which have become comparable to the evolved feature sizes in modern microelectronics. In Figure 1.9, the scaling trends of some key feature sizes in electronics set by the International Technology Roadmap for Semiconductors (ITRS)[62] and the Intel® corporation [63] are presented along with some typical feature sizes in nature for comparison. The applicability of LET in radiology is out of the scope of this work. In the following section, the dependence of feature size and projectile energy on the mean energy deposition, in electronics point of view, is discussed.

1.3.4 Spatially Restricted LET

The energy lost by an ion, discussed above, is transferred to kinetic (T_{kin}) and potential (T_{pot}) energy of the target electrons. In order to contemplate the spatial distribution of the deposited energy, first, the energy transfer to the target electrons needs to be defined. By rewriting Equation (2.7), the differential cross section for Coulomb scattering becomes

1.3.5 Energy Loss Straggling

Energy loss of an ion in a single collision with the target electron is a stochastic event considered to follow the Poisson statistics. The expressions for the stopping force, discussed above, give only the average energy loss per unit length. In relatively large volumes, this applies, but when the volume decreases, also the randomness in the energy loss, or deposition, becomes significant, in addition to the spatially restricted LET discussed above.

A simple demonstration of the dependence of the SV size on the fluctuation in the deposited energy is presented in Figure 1.11. Here, results from Geant4-simulations [92, 93] for energy deposition of 1-MeV α-particles passing through silicon targets with different thicknesses are presented. The simulations were made by "bombarding" slabs of silicon (cf. Figure 1.6) with α-particles. The lateral dimensions (W and H) in the slabs were much larger than the thickness ($\triangle x$). Thus, from Equation (39), the mean chord length becomes $l = 2\triangle x$. The effect of the spatially restricted energy deposition is considered to be less significant in this case. It is seen in Figure 1.10 that, for 1 MeV α-particles, from the total energy loss $\sim$90% is deposited within 20 nm from the ion's trajectory. Thus, we can assume in an SV with the mean

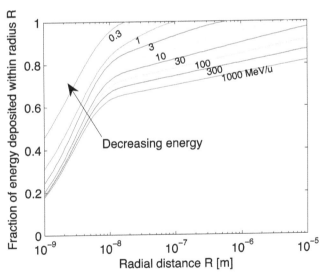

Figure 1.10 Fraction of deposited energy within radius from the ion track as a function of radial distance for ion velocities (or energies) 0.3, 1, 3, 10, 30, 100, 300 and 1000 MeV/u, from top to bottom, respectively. Figure taken from Ref. [4].

chord length of >20 nm (corresponding to the 10 nm slab), the deposited energy is more than this percentage of the average energy loss.

The shift in the spectra in Figure 1.11 to higher energies for thicker targets originates from the decrease in the ion energy (thus increase in the stopping force) along the ion's path. This increases the total deposited energy in the thicker targets compared to the average energy loss for 1-MeV α-particles given by the SRIM-calculations, ~305 keV/μm. Also, one should bear in mind that the descriptions for stopping in the SRIM and the Geant4 are slightly different, thus possibly yielding different average energy loss values. The physical descriptions for the nuclear reactions, discussed above, were omitted in these simulations.

The most distinct feature in this graph is the broadening of the spectra due to the straggling. The dependence of the target thickness on the broadening is clearly observed. In thin targets, the relative fluctuation in the deposited energy around the average value increases with decreasing thickness. For example, in Figure 1.11, for the 10 nm silicon target, some of the α-particles deposit even twice the average energy.

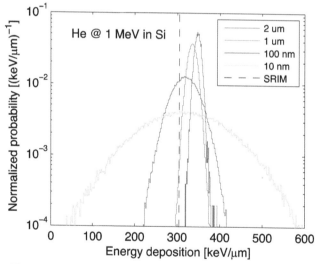

Figure 1.11   The spectra of deposited energy from Geant4-simulations for 1 MeV α-particles traversing silicon targets with thicknesses of 1 μm, 2 μm, 100 nm and 10 nm. The electronic stopping force value for 1 MeV α-particles, taken from SRIM, is plotted with dashed line, is ~305 keV/μm. Figure taken from Ref. [4].

1.3.6 Applicability of LET

By combining the above-mentioned aspects with the concept of LET, different regions governed by these mechanisms can be separated in the phase space determined by the ion's velocity and SV's mean chord length. These regions are depicted in Figure 1.12, where the same definitions are used as in Refs. [53, 55]. The limits drawn in the graph represent the so-called 10% thresholds, where the individual effects (spatial restriction and/or straggling) are causing more the 10% variation in the energy deposition defined by the electronic stopping. Here, the limits are calculated for protons, α-particles and N-ions to demonstrate the dependence of the ion specie on the effects. The different regions in the graph are explained in the following list.

For protons and heavier ions, the LET can be considered to describe the energy deposition with sufficient accuracy.

The effect of straggling contributes to the energy deposition for protons, but for heavier ions, the use of LET is still valid.

The straggling for the α-particles becomes significant.

The spatial restriction for the energy deposition comes into play for all ions, and the straggling for N-ions still plays a minor role.

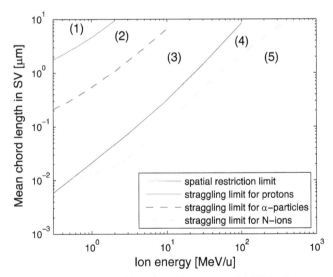

Figure 1.12 Regions of mean chord length and ion energy (MeV/u) where parameters LET, straggling and spatial restriction affect the energy deposition. The different regions (1)–(5) are explained in the text. Figure taken from Ref. [4].

For N-ions, all three concepts, the LET, its spatial restriction and the straggling, need to be taken into account when considering the energy deposition.

The above discussion shows that when the technology is approaching the nanoscale, the conventional description of LET becomes increasingly insufficient to characterize the phenomena in the electronics. Thus, more sophisticated models would be needed. At this point, there is no simple concept to replace the LET as a metric used in the field of radiation effects in electronics.

1.3.7 Prediction Tools for Stopping Force

Because no consensus has been reached on the alternatives among the community, the LET concept has yet remained in use in the characterization of the radiation effects. Moreover, as the concepts of the spatially restricted LET and the straggling require information on the SV(s) in the studied structures, which is usually unavailable, the values of electronic stopping force are used as the LET. For estimating the electronic (and in some cases also nuclear) stopping for a given ion in a given target, there is a variety of tools available, both semi-empirical and theoretical, in addition to tabulations of experimental data. A collection of LET prediction tools and data tables are presented in Table 1.1 with a brief introduction. From the listed tools, the SRIM code has received the widest acceptance among the users, and at least in the radiation effects community, it is the primary repository of stopping force values. This is due to its user-friendly interface, and in average, if all ion–target combinations are accounted for, its accuracy is fairly good.

In addition to these tools listed in Table 1.1, the Geant4 simulation toolkit [74] has become increasingly into use in predicting radiation effects in electronics. The disadvantage in the Geant4 calculations is the relatively high threshold for deployment. On the other hand, the Geant4 is a powerful tool to investigate radiation effects in complex volumes. There is also a Geant4-based Monte-Carlo Radiative Energy Deposition (MRED) tool developed in Vanderbilt University, USA, which is available online [10]. This tool enables estimations for the particle-induced energy deposition in complex geometries via a user-friendly interface. For this tool, user registration is required, whereas the Geant4 is fully open source.

All the above mentioned are estimation tools (except the data tables), and they provide estimations for the stopping force. Each of them have

Table 1.1 Stopping force prediction tools.

2 Source	Type	Description
SRIM [64]	Semi-empirical	The most well-known of them all. Based on the work of Ziegler et al. described in Ref. [94]. Over the years, numerous updates have been made in the calculations. Last major update was in 2003. This code and its user interface are very versatile and user-friendly. It enables calculations for a wide energy range with huge variety of ion-target combinations with a reasonable accuracy. Software is freely available at http://www.srim.org.
LET Calculator [65, 66]	Semi-empirical	Another relatively widely used tool developed by Zajic and Thieberger in Brookhaven National Laboratory. Based on an earlier version of SRIM with parametrization to experimental data measured by the developers. Available online at http://tvdg10.phy.bnl.gov/let.html.
ECIF Cocktail Calculator [67]	Semi-empirical	Based on parametrization of modified Bohr's classical stopping theory [68]. Web interface can be found at http://research.jyu.fi/radef/ECIFcalc/dedx.html.
PASS [69]	Theory	Based on the work by Sigmund and Schinner [70]. The calculations are fully based on fundamental physics without any parameterization to the experimental data. Stopping force values are available upon request from the model developers.
CaSP[71]	Theory	Convolution approximation for Swift Particles developed by Grande and Schiwietz and described in Ref. [101, and references therein]. Available online at http://www.casp-program.org/.
Paul's database [72]	Data table	Extensive tabulation of published experimental stopping force data maintained by Helmut Paul. Available online at https://www-nds.iaea.org/stopping/.
Hubert tables [73]	Data table	Tabulation of experimental stopping force values in the energy range from 2.5 to 500 MeV/u. Not as comprehensive as Paul's database above.

their pros and cons, and in cases where there is no experimental data to compare, it is in the user's judgement whether or not to rely on the values given by the predictor. Discussions have aroused among the radiation effects community about the differences in the estimated stopping force values between the tools especially for heavy ions like xenon [75]. This is due to the extrapolation of the parameters based on experimental data in other ion–target combinations.

1.3.8 Cumulative Effect: Total Ionizing Dose and Displacement Damage

Cumulative stress of radiation may cause gradual changes in the characteristic properties of electronics, such as threshold voltage shifts and decrease in the minority carrier lifetimes. There are two subdivisions for the cumulative effects: Total Ionizing Dose (TID) and Displacement Damage Dose (DDD). They are related to the concepts of the LET and the NIEL, respectively.

The TID effects are governed by those radiation-induced charge carriers, which have survived the recombination and are not swept away by the electric fields. Because the electron mobilities, especially in dielectrics, are higher than those of holes, typically the trapped charges are the holes. Moreover, the effect is more pronounced in dielectrics than in semiconductors. Although, in case of interface trapped charge, it is located in the semiconductor side of a dielectric–semiconductor interface. The TID effects can occur in both Metal–Oxide–Semiconductor (MOS) devices and bipolar devices. For more detailed information on TID effects, the reader is referred to Refs. [76, 77].

The DDD effects in electronics are induced by the NIEL associated with particle radiation (see Section 1.3.5). The atoms may get knocked out from their lattice site due to an ion hit, creating a vacancy–interstitial pair, called a Frenkel defect or pair (FP). Typically, the reduction in current gain of bipolar transistors or dark currents in CCDs are attributed to these defects. This work mainly focusses on the ionizing particle radiation, and the effects of NIEL or DDD are not discussed further.

1.3.9 Single Event Effects

The Single Event Effects (SEE) are referred to as the prompt response of electronics to ionization event, induced by a single energetic charged particle, as depicted in Figure 1.13. As discussed above, charged particles are capable of ionizing matter, i.e. generating e–h pairs in semiconductors and dielectrics. Typically, SEE occur when an ion strikes a reverse-biased PN junction in an off-state device, creating a transient, unwanted flow of current which may be sufficient to cause errors and/or physical damage. SEEs can be divided into two groups: non-destructive soft errors, and destructive hard errors. In the following sections, these two groups are briefly introduced.

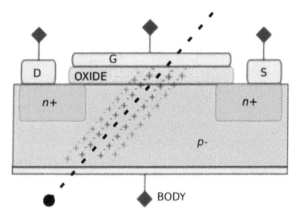

Figure 1.13 A schematic view of an ion hit in an NMOS.

1.3.10 Soft Errors

When ion induces a temporary disturbance in an electronic circuit, which can be fixed by e.g. reprogramming the device, the effect is considered as a soft error. The main types of soft errors and their definitions according to Refs. [78, 79] are listed below.

Single-Event Upset (SEU) is an event where a memory bit (or bits) is (are) flipped, from 0 to 1 or vice versa. Typically, only one bit is affected at the time and the effect can also be called Single Bit Upset (SBU). In case of corruption of multiple bits, the event is referred to as Multiple Bit Upset (MBU), which can be a concern in highly scaled memories or with ion hits at high grazing angles. The basic principle of an SEU in an SRAM cell based on Complementary Metal–Oxide–Semiconductor (CMOS) technology is discussed below.

Single-Event Transient (SET) is a progressive disturbance in combinatorial logic systems caused by a single ion hit. The ramifications of an SET are dependent on e.g. the operation frequency of the circuit. SET may turn into an SEU, if it gets latched. In case of analogue devices (e.g. operational amplifiers, comparators and voltage regulators), the transient disturbance due to an ion hit is called Analog Single Event Transient (ASET). The faulty signal caused by an ASET can propagate in an integrated circuit and lead to significant anomalies, such as data corruption or system failure [22, and references therein].

Single-Event Functional Interrupt (SEFI) results in a loss of device functionality. After occurrence of SEFI, the malfunction can be fixed without power cycling the device. SEFIs are typically associated with SEUs in a control bit or in register.

Because the SEU is the "oldest" SEE-type, being observed in the first studies [13], it is chosen here as an example. The anatomy of an SEU in a typical SRAM cell is illustrated next.

The mitigation of these above-mentioned soft errors has become more important, not only in space electronics but also in general commercial electronics, due to the technological evolution [81]. The mitigation techniques can be applied either in circuit or software level. The reader is referred to Ref. [82] for a detailed discussion on different mitigation methods for soft errors in modern microelectronics, and they are also discussed later in this book.

1.3.11 Hard Errors

In some cases, particle-induced "cloud" of e–h pairs may generate current peak in a device, which may initiate high currents, leading to a destructive failure. These kinds of irreversible hard errors are typically less probable to occur than the soft errors discussed above. Of course, their weight in radiation reliability of electronics is higher because once they occur, the device is partially or totally out of service. The typical destructive SEE types are listed below with short descriptions after Refs. [78, 79, 83]:

Single-Event Latch-Up (SEL) is potentially a destructive state in a device, where hit of a single ion creates conductive path between the device power supply and ground. The current through this path is stopped only by shutting down the power supply. The SEL may destroy the device if the current from the power supply is not limited and/or the power cycling is not performed fast enough after the current increase.

Single-Event Gate Rupture (SEGR) is a breakdown of gate oxide in MOS devices, attributed to ion-induced conductive path. The excessive current through the dielectric leads to material meltdown via thermal runaway. The basic physical mechanisms underlying SEGR are yet unknown, mainly because the rapid nature of the event impedes accurate measurement of the current spikes. Only some qualitative or semi-empirical models exist for SEGR prediction.

Single-Event Burnout (SEB) is a failure that can be observed typically in power devices (MOSFET or bipolar). In the SEB event, there is a highly conductive path created in lightly doped epitaxial layer of the device, which leads to excessive current, and ultimately to thermal runaway with a permanent damage. In many cases, SEB and SEGR occur simultaneously in power MOSFETs.

Mitigation of SEL and SEB is possible by limiting the supply current of the device. The SEL is also possible to avoid by using Silicon-On-Insulator (SOI) technology [84]. However, SEGR is found to be impossible to avoid once the threshold conditions (oxide electric field and energy deposition density) are exceeded [83].

1.4 Effect of Radiation on Memory Devices

Ionizing radiation affects many types of electronic devices and, among them, memories are the most concerned by soft errors. According to the ITRS roadmap [Semiconductor Industry Association (SIA), "International Technology Roadmap for Semiconductors (ITRS); http://www.itrs.net], electronic memories are the dominant type of devices in embedded systems, representing the largest quota in Systems on Chip (SoC) area. Due to their rather simple architecture, compatibility with standard process and performances, memories are used in most computing systems. On the other hand, the inner characteristics of this type of device make them one of the main sources of errors in SoCs [85]. Moreover, the storage capability, that is, the main feature of all memories, makes them capable of storing radiation-induced errors, differently from combinational circuits where soft errors are less frequent and more difficult to detect.

1.4.1 Structure of a Memory

In order to ease the understanding of the interaction between ionizing particles and memories, it is useful to introduce the typical structure of a memory device. As depicted in Figure 1.14, memories are mainly composed of the memory cell array, where the data are physically stored, and peripheral circuitry which allows cell selection, data reading/writing, buffering and synchronization. Besides the memory cells, all the elements of peripheral circuitry may also be impacted by the radiation environment.

The memory array is not always composed of a single rectangular cell matrix, as seen in Figure 1.14, but it can also present more complex configurations such as the butterfly, with two cell arrays divided by a single row decoder [85], as depicted in Figure 1.15. This configuration reduces the word line length by half. At the same time, the bits of each word are split into two sub-arrays, with the row decoder physically dividing them.

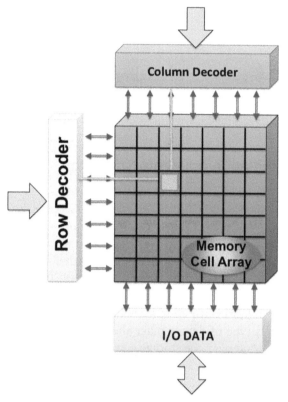

Figure 1.14   Basic structure of a memory.

Simple or multiple arrays can be further organized into blocks, allowing the reduction of the size of bit lines and word lines, making the cell access (read/write) more effective and reliable, since the shorter wiring reduces delays. Another important feature of cell array is related to the size and location of the bits belonging to a memory word, which is generally made of between 8 and 64 bits. The cells corresponding to these bits are usually topologically distant, since interleaving schemes are applied. These schemes place all the bits belonging to the same word in different electric blocks, and the access to these bits during read/write operations are made through multiplexing structures. An example interleaving is given in Figure 1.16 that refers to a section of SRAM array, in which the cells are organized in 4×4 blocks.

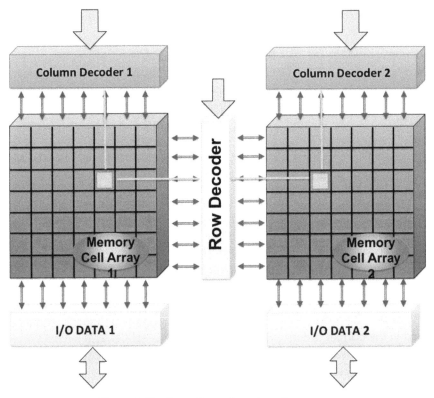

Figure 1.15 Butterfly configuration of a memory.

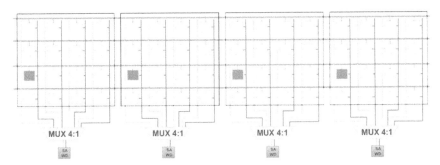

Figure 1.16 Scheme of a section of an SRAM array organized in 4×4 blocks.

The selection of a given word for a write/read access is done as follows:

- The word line that selects the given word is activated;
- Only one column per block is selected;

- The selected cells are connected to one read/write sense amplifier/write driver (SA/WD) unit through a 4:1 multiplexer.

In this configuration, the bits belonging to the same word are never topologically adjacent, but they are placed in separate blocks, with a minimum distance of three cells. In the case of blocks composed of 16×16 cells, the minimum distance between two cells of the same word will be 15 cells.

In addition to interleaving schemes, scrambling techniques are also implemented in the cell array, thus contiguous values of addresses are generally not physically adjacent.

The main peripheral circuits are listed below:

- The **address decoders**, which allows selecting the row and column where to read/write;
- The **sense amplifiers**, which sense the state of the memory cells is in order to convert it in binary form;
- The input/output (I/O) **buffers**, where the content of the fetched data is retrieved prior to latching the data on the I/O pins or wrapper;
- The **synchronization circuitry** ensuring that each action involved in the read/write action is performed within predetermined time constraints and in the correct order;
- The **write drivers** consisting of the logic elements used to force the logic values in the cells;
- The power and ground **grids** that feed the memory array as well as the peripheral circuitry;
- The **power switches** that allow switching ON/OFF parts of the circuit, especially in the case of low-power (LP) devices. For example, in LP SRAMs, power switches allow feeding the cells with different supply voltage values: low voltage ($\sim$0.74 V) to reduce leakage currents (and hence device power consumption), while still maintaining data retention; high voltage (> 1.1 V) to allow read/write access without destroying the cell content during a read operation or failing the write of a cell due to the reduced write noise margin.

Depending on the considered type of memory and technology, the electrical and physical function of the device may change and require the use of specific circuitry and structures, such as charge pumps in flash memories, and a microcontroller and data buffers to control the I/O data exchange in large storage devices. In the following chapters, when needed, these specific structures will be presented and detailed, when needed to explain the introduced hardening techniques.

1.4.2 Classification and Fault Mechanisms in Memories

All malfunctions (related or not to ionizing particles) affecting any subcircuit of a memory are systematically mapped as error(s) in the memory array. These errors are in part actual bit flips and in part false bit flips, as in the case of read failures, in which although the selected cells store the correct data, the output data are incorrect. Based on these premises, the soft errors that affect memories can be classified as follows:

- Single Bit Upsets (SBUs) are generated when a particle-induced current flips the information stored in one memory cell (or a latch or Flip Flop in other types of circuit).
- Multiple Cell Upsets (MCUs) are special cases during which a single particle is responsible for the upset of more than one-bit cells. This phenomenon can be the result of a single particle nuclear reaction generating multiple recoils that travel at different directions deposing charge in multiple sensitive nodes, or the result of a single ion deposing sufficient amount of charge to upset two or more neighbouring cells [86].
- A special case of MCU is the Multiple Bit Upset (MBU), where more than one bit of the same word is corrupted. The application of interleaving schemes, introduced in the previous section, in the cell array (bit-cells of other words placed between consecutive bit-cells of the same word) makes MBUs unlikely to occur. Nevertheless, with the miniaturization of transistors - and consequently of bit cells, the probability of MBU appearance has increased in the recent years. This is mainly the result of the reduction of the distance between sensitive nodes, and the comparatively long range achieved by high-energy ions in silicon.

In order to show how ionizing particles may induce soft errors in memories, two examples of fault mechanism are introduced here, considering a single event affecting a memory cell (SRAM) and a peripheral circuit (address decoder). As introduced above, the resulting errors are mapped in the memory array, thus interpretable as error(s) in read bits.

Example 1: Particle hitting an SRAM core cell.

Considering the typical case of SRAM devices and the mechanism of indirect ionization, the fault mechanism can be explained as follows. The triggering event may be either:

- An energetic charged particle goes through the substrate of the target device, and ionizes target atoms along its path;

- An energetic neutron or ion impinges on a target atom within the device. Depending on the impinging particle energy and target atom isotope, a nuclear reaction can take place, possibly generating heavy ions as reaction products. These ions, retaining the momentum of the original impinging particle, will recoil in the silicon and ionize target atoms along their path;

In either case, a track of electron–hole pairs is created in the silicon. If such a track is generated close to a reverse-biased junction of an SRAM cell, the carriers will drift across the depletion region, creating a current spike [87]. The fast collection of carriers is completed within a few hundreds of picoseconds, while the collection takes place over a few hundreds of nanoseconds, until the carriers start to diffuse. Figure 1.17 illustrates the failure mechanism in an SRAM cell.

The parasitic current that is generated by the secondary ions occurring from the particle collision to the reverse biased junction of the drain of the NMOS transistor Mn4, is represented at the SRAM cell schematic with a current source. This current source discharges node SB (formerly at logic "1", VDD), enabling the activation of transistor Mp3, thus leading to the complete flipping of the cell state (node S at logic "1", node SB at logic "0"). This event can be the result of a neutron or high-energy proton impinging close to the junction. Alpha particles, heavy ions and low-energy protons, on the other hand, induce the parasitic current by traversing directly the silicon, leaving the electron–hole pair track on their path, resulting in a direct ionization mechanism.

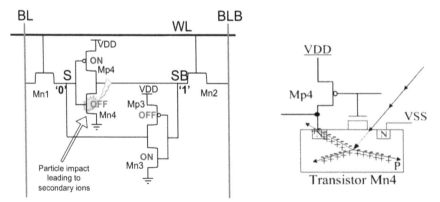

Figure 1.17 SRAM cell fault model. A particle induces a nuclear reaction, which generates secondary ions. The ions create electron–hole pairs, which are drifted to the depletion region of the reverse biased junction of the drain of transistor Mn4.

When considering other types of memories, the fault mechanisms responsible for data corruption may be different, and will be detailed in the following chapters. For example, in a flash memory, the memory cell is composed of a floating-gate transistor, and the cell's stored logic value depends on the transistor's threshold voltage, hence, on the amount of charge stored in the floating gate. In this case, impinging ionizing particles may lead to data corruption by modifying this charge level, and/or cell state sensing through the accumulation of parasitic trapped charges in the transistor dielectric materials.

Example 2: Particle hitting an SRAM address pre-decoder.

This example shows how MBUs within the same word can be induced by temporary malfunction of the address decoder, as opposed to classic MCUs caused by a particle hitting the memory cell array. Figure 1.18 depicts the schematic of an address pre-decoder, which is composed of two stages. The first stage (on the left-hand side in the schematic) has synchronization function, with WLEN as a synchronization signal. The second stage (on the right-hand side) executes the decoding and amplification of the word/bit lines selection signals. For this example, SPICE simulations were performed considering a single ionizing particle hitting the first stage of the decoder, by using the charge injection model described in [88].

The injected charge induces an address decoder malfunction that provokes a double word line selection during a write operation. During a correct

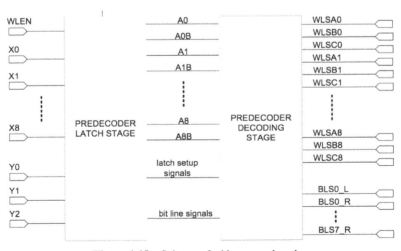

Figure 1.18   Scheme of address pre-decoder.

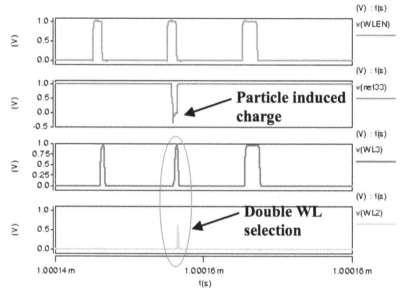

Figure 1.19   Double word line selection during write operation.

address function, only a single word line (and thus a single word) selection per access is possible. Consequently, in case of faulty double word selection (due, in this case, to a particle hit) the data to be written are stored in the correct word (aggressing word) as well as in another one (victim word), potentially producing, in the latter, multiple bit flips. Figure 1.19 depicts this faulty behaviour: the word line signals WL2 (victim) and WL3 (aggressor) are activated at the same time. Considering an 8-bit victim word storing 00100101 before the faulty write, and the new stored data is 11111111; this introduces a 5-bit MBU (5-bit flips within the word).

Since address decoders are very similar, independently of the type of the memory, this same fault mechanism can be observed in several types of memory devices.

1.4.3 Memory Accelerated Tests

Functional and radiation-induced faults in an SRAM are either generated in the memory cell array, or are mapped in it. In other words, the error can be either generated in the core cell(s) and be observable with the cell read access, or be generated in the peripheral circuits and result in read/write failures or an access to wrong memory locations. These failures are next observed with read accesses that only apparently reveal cell bit flips. For example, if during a read access, the synchronization circuitry is affected by a particle, the sensing time can be reduced. Consequently, the voltage difference between the two cell nodes to be sensed is reduced, and the value returned by the read operation is random, and thus potentially faulty. For an external observer, the error appears like a cell bit flip, although the cell still stores the correct value. In this case, a second read access will reveal whether the error is due to an actual bit flip, or to a failure in the access circuitry.

1.4.4 Test Methods

Being the main process used for the qualification of electronic components in applications concerned with radiation effects, the accelerated test is a complex procedure that involves several parameters related to the device under test (DUT) and the environment of exposure. When it comes to evaluating a component's radiation sensitivity, the most important metrics are its cross-section (the area of the device that is sensitive to particle hits) and the minimum linear energy transfer (LET) threshold for impinging particles to cause an error. Although SRAMs are relatively simple components compared to Systems-On-Chip (SoCs) or microprocessors, their testing can hide several aspects that can affect their sensitivity estimation and need to be taken into consideration. Directives in the form of standards are given such as the JEDEC [89, 90], which provide fundamental guidelines for the testing of electronic devices. The guidelines that concern memories are mainly focused on static mode testing and display some complementary notes on dynamic mode testing. In the following paragraphs, the details and importance of each testing mode are analyzed, and various methods related to the dynamic mode test are presented.

Considering a simple level of abstraction, memory operation can be classified into two distinct modes: static mode (retention) and dynamic mode (read/write access). This abstraction is based on the major differences in the behaviour of the memory cells and controlling logic under these two modes. In static mode, memory cells perform data storage: the access transistors of the cell are OFF, and no interference on the cell is operated by normal function. When the cell is accessed for a read operation, the access transistors are activated, the stored logic value is determined by a sense amplifier, which returns either a logic "0" or a logic "1". During a write operation, with the access transistor activated, the bit line(s) (properly set) force either the preservation or the flipping of the stored information in the cell. The activation of the peripheral circuitry of the memory (address decoders, write drivers, pre-charge circuitry, etc.), the electric weakening of the cell (due to the read access in some types of memories like SRAMs) and other phenomena, typical of dynamic operation mode, do not occur in static operation mode. Conversely, when in static mode, the cell is in retention, which for many cases (especially for low-power SRAMs) means that the voltage in the memory array approaches its threshold limit reducing the static noise margin of the cell. These differences make the cells more or less susceptible to Single Event Upsets (SEUs) such as SBUs, MCUs, MBUs and Single Event Functional Interrupts (SEFIs). Consequently, it is essential to differentiate the modes of testing, depending on the level of SEU sensitivity of the memory in Static and Dynamic modes.

1.4.5 Static Mode Testing

Static mode testing is one of the most fundamental testing methods for the vast majority of memories, and especially SRAMs. It is described as the key test method by various standards, and it requires the writing of the memory with a known data pattern prior the exposure to the radiation source (natural/real-time, accelerated, etc.). Following the exposure of the memory under radiation for the time or scheduled fluence, the memory is read back to check for possible bit-flips. The data background must be stored on the memory prior to the irradiation, in order prevent "dynamic" errors during the initializing (writing) process.

Static mode testing has a major benefit, which is the rather stable response of the memory in terms of sensitivity. Since the only part of the memory that is affected is the cell array and not the periphery, all errors are classified into two major categories, which are easy to distinguish and define the error probability, as opposed to the dynamic mode testing that has several different sources and types of errors. The two major categories of errors are SBUs and MCUs. An SBU occurs when a particle impinging the memory cell array induces a parasitic charge through either direct ionization (heavy ions, alphas, low-energy protons) or indirect ionization (neutrons, protons), and this charge flips the bit stored inside one cell. An MCU, on the other hand, is the result of the same phenomenon, with the difference that the parasitic charge affects multiple cells that are topologically adjacent in the memory array.

During static mode testing, the choice of the data pattern depends on the type of test and the depth of analysis that needs to be achieved. Typical sets of data backgrounds are the following ones: solid 1 (all "1"s), solid 0 (all "0"s), checkerboard ("10101010") and inverse checkerboard ("01010101"). Comparison between the solid 1 and solid 0 data patterns can reveal the relation of the stored information (logic "1" or "0") with the sensitivity of the cell, as a result of intra-die variations, or resistive/bridging defects within the cells or the memory array. Checkerboard and inverse checkerboard may reveal differences in the overall response of the memory array, as well as differences regarding the shapes and sizes of MCUs, as shown in [91].

1.4.6 Dynamic Mode Testing

Although in the various standards describing radiation testing a basic methodology is provided for the dynamic mode testing of memory devices, no detailed guidelines are given, and moreover, no special cases are taken regarding the memory under study and its respective technology. Such guidelines cannot cover all types of memory devices when operating in dynamic mode, because each type of device has its own sensitivity that also depends on the access pattern (sequence of read/write operations). A certain access pattern may stimulate (stress) one device and leave another one without any stress effect. As mentioned above, when the memory operates under static mode, only the cell array is susceptible to SEEs, while when in dynamic mode, other types of SEUs may occur besides typical SBUs and MCUs such as SEFIs and SELs: additional regions of the memory are activated, and thus become vulnerable to impinging particles. For example, the injection of a parasitic current in the address decoder during a write operation may result in storing data in wrong memory locations, as shown in the example given in Section 1.4.2.

Finding the proper testing scheme to stimulate a memory device under its typical, best-case or worst-case scenarios is not evident when operating in dynamic mode. March algorithms are currently used in the manufacturing process because they are capable of detecting faults in memory devices, such as bridging faults, stuck-at faults and coupling faults among others. They are preferred for their efficiency and low complexity. Depending on the type of memory, different algorithms are preferred based on their efficacy, thus, some algorithms are expected to be applied for SRAMs and different ones for other types of memories such as DRAMs and FLASH. Since March algorithms are able to stimulate the memory devices in order to reveal manufacturing defects, providing a suitable functional test, they become the perfect candidate for radiation dynamic testing, as they can be tuned to operate the typical as well as the worst-case scenarios of the memory's operation.

Table 1.2   Marching test algorithms

Name	Test
March C−	{↑(w0);↑(r0,w1);(r1,w0); ↓(r0,w1); ↓(r1,w0); ↑(r0)}
Mats+	{↑(w0);↑(r0w1);↓(r1w0)}
mMats+	{↑(r0w1);↑(r1w0)}
Dynamic Stress	{↑(r1,w0,r0,r0,r0,r0,r0); ↑(r0,w1,r1,r1,r1,r1,r1); ↑(r1,w0,r0,r0,r0,r0,r0); ↓(r0,w1,r1,r1,r1,r1,r1); ↓(r1,w0,r0,r0,r0,r0,r0); ↑(r0,w1,r1,r1,r1,r1,r1)}
Dynamic Classic	{↑(w0);↑(r0);↑(w1);↓(r1)}

Table 1.2 shows a list of the March algorithms introduced in various works by the authors, such as the Dynamic Stress (March DS), the March C- and the Dynamic Classic among others.

Each March algorithm is made of several elements. Each element entails a series of operations (read or write). Each element has a given addressing order of execution: the arrow at the beginning of each element indicates the order of the addressing, i.e. from the highest memory address to the lowest ($\downarrow$), or from the lowest to highest ($\uparrow$). The operations of each element are all applied to each address location (word), before proceeding to the next address. This means that all the operations of the element are applied to a single word, before the address counter of the tester is increased, or decreased, depending on the direction ($\uparrow$ or $\downarrow$) of the element. A semicolon separates the elements of the algorithms and each element has its operations enclosed in brackets. For example, the fourth element of March C- $\downarrow$(r0, w1), has an addressing scheme starting from the uppermost address towards the lower ones, and applies a "read '0'" followed by a "write '1'" operation into all the memory locations. This means that the tester reads a single word and expects to receive a zero pattern, after which it writes all the bits of the same word with logic "1". Once these two operations are finished, the tester proceeds to the next word, and applies the same pattern.

For all the algorithms in Table 1.2, the data background stored and read from the memory is either a solid "0" or a solid "1" and the swapping from "0" to "1" (or the opposite) stresses (by increasing the switching activity) some peripheral circuitry of the memory such as the I/O buffers. Other parts of the memory can be stressed as well such as the address decoders by using the specific techniques. For example, the address decoder can be stressed by applying an addressing sequence in which most of the bits change at each access, strongly enhancing the switching activity of the address buffer and decoders, as explained in [85].

Depending on the given March algorithm, a strong stress factor can be induced to the memory such as the cases of March Dynamic Stress (DS) for SRAMs, while others like the Classic are less stressful, when not properly modified. As demonstrated in [92], the algorithm March DS induces a stress to the SRAM cell by applying sequences of read operations in the same locations. When sequentially made, the read operations reduce progressively the static noise margin of the cell by degrading the voltage level of the nodes. As explained in [93], the March DS algorithm, which is proven for the SRAMs to be amongst the most stressing algorithm, has the opposite impact on FRAMs. Considering this example, it becomes clear that each type of memory has its own particularities, which need to be well analyzed before structuring a test methodology.

In order to enhance or reduce furthermore the stress of the memory array and the periphery, some additional properties of March test algorithms are detailed in [94], such as Fast Row; Fast Column; Random Addressing; Adjacent (Gray) Addressing and Inverse Gray Addressing. These complementary techniques to already existing March algorithms affect the level of electric stress induced to the address decoders, the data buffers and other peripheral circuitry by playing on the address order and the data background.

1.5 Radiation Hardness Assurance Testing

There are various particle radiation sources present at the ground level and in space. The radiation hardness of electronics operating in harsh radiation environments, such as nuclear power plants, accelerators or in space, needs to be assured either by manufacturing them with radiation hardness in mind, or by testing them. In earlier days, there were dedicated manufacturers for rad-hard components, which were competitive in performance with commercial products. When the Cold War ended and the rivalry over space supremacy cooled down, the production of rad-hard electronics was no longer a prosperous business. Only a few manufacturers remained in the field, none of them exclusively on rad-hard products. Because of high production costs, nowadays the rad-hard industry is mainly focused on parts with high reliability requirements. In addition to lower prices, the performance of commercial electronics (COTS, commercial-of-the-shelf) is typically much higher than that of rad-hard devices. Thus, COTS devices are often favoured in space projects due to superior performance and low cost; however, the drawback is their lack of being space qualified and unknown radiation performance.

As said, to ensure the radiation hardness of an electronic component which was not manufactured rad-hard, the component needs to be tested under radiation. There are some requirements for these Radiation Hardness Assurance (RHA) tests, set by the radiation effects community. Some of the most commonly used specifications are presented in Refs. [78, 95–97]. These documents determine the framework within which RHA tests typically need to be performed. As the aforementioned particle radiation environments are dominated by light particles (i.e. protons and electrons), the total dose issue is the first to be considered. Due to the focus of this work, which is on particle radiation and particularly on heavy ions, the total dose testing is not discussed further. Instead, an introduction on the basics for RHA testing for SEEs is given.

The radiation environment is usually represented by its LET spectrum, which is considered to represent a particle's ionizing power. This can be done by first defining individual particle spectra (see Figure 1.4) in terms of their LET and merging them together. The result is a single spectrum representing the given environment. As an example, four LET spectra are presented in Figure 1.20, where radiation environments in two different regions are estimated during solar maximum and minimum (the data are taken from Refs. [10, 99]). Because of the presence of heavy ions (or high LETs) in these environments, not only the total dose effects are a concern, but the electronics need to be determined also for their susceptibility to SEE. In principle, SEEs can be produced by any source emitting particle radiation, but the particles must have sufficient energy to penetrate into the device all the way down to the silicon, and be able to generate enough ionization. In addition, the used radiation needs to mimic the effects in the real operational environment.

In laboratory, radiation sources, such as Ra-226 ($\sim$100% α-emitter, with some gamma radiation) and Cf-252 (α-emitter, with 3% of spontaneous fission, producing also neutrons) [100] are convenient for this purpose because of their relatively low price and ease of use. The disadvantage in these sources is the limited ion energies and species as well as the limited particle flux. Furthermore, there may be more than one particle type, emitted at several different energies from the source. Also, radiation type may be multifaceted, e.g. including also gamma or neutron radiation, which limits the usefulness of these sources even more. Due to these limitations, the use of radioisotope sources is prohibited in the official RHA tests by the specifications [78, 95]. These sources are typically used only for trial runs of the setup for the

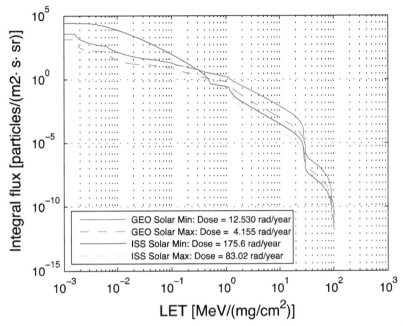

Figure 1.20 Integral flux of galactic cosmic ray particles during solar minimum (solid) and solar maximum (dashed) as a function of Linear Energy Transfer in GEO or near Earth interplanetary space (blue and green) and International Space Station orbit (red, cyan). Data taken from [10]. Figure taken from Ref. [4].

RHA testing. The specifications for RHA testing for SEEs require the use of particle accelerators.

The particle accelerators are multi-purpose tools, which in the past have been used mainly in basic nuclear physics studies. In recent years, accelerators have been increasingly used for applied research, such as RHA tests. Next, a short introduction is given on the basic principle of SEU testing using heavy ions.

As discussed above, SEUs are the most prominent error type in electronics. SEU testing of e.g. memory devices is done by irradiating a memory with certain ions (each ion species yielding a maximum LET in a given target material), at fluences ranging from 10^6 to 10^7cm^{-2} depending on the device sensitivity [78, 95]. Typically, above the sensitivity threshold, more than 100 upsets are required to obtain adequate statistics. The required ion fluxes are from 10^2 to $10^5 \text{cm}^{-2}\text{s}^{-1}$. In a typical test procedure after the ion exposure, the memory is read and the number of bit flips (SEUs), N_{err}, are determined. From this, the SEU cross-section per bit can be calculated by using the equation:

$$\sigma_{SEU} = \frac{N_{err}}{n_{bit} \cdot \Phi},$$ (1.14)

where n_{bit} is the number of bits in the tested memory and Φ is the ion fluence given in cm^{-2}. By changing the LET (using different ions), the characteristic SEU sensitivity of a device can be obtained. Moreover, the LET can be varied by tilting. This introduces a concept of effective LET which is defined as:

$$LET_{eff} = \frac{LET}{\cos(\theta)},$$ (1.15)

where θ is the angle of incidence for the impinging ion, and LET is the electronic stopping force of the ion. By definition, $\theta = 0°$ when the ion trajectory is perpendicular to the device surface. The applicability of the effective LET has also been questioned over the years, along with the concept of LET in general (see discussion above). No further comments will be made on the suitability of the effective LET concept here. One should also note that under tilted conditions the ion fluence, Φ, should be replaced by the effective fluence:

$$\Phi_{eff} = \Phi \cdot \cos(\theta),$$ (1.16)

In Figure 1.21 are SEU cross-section data for an SRAM plotted against (effective) LET. This is a typical way to characterize the radiation susceptibility of electronics. In this plot are data measured in three different facilities: the Heavy Ion Facility (HIF) in Louvain-la-Neuve, Belgium [91], the Radiation Effects Facility (RADEF) in the University of Jyväskylä [102] and the Radiation Effects Facility in the Texas A&M University, USA [93].

This kind of characteristic plot can be used to estimate the SEU rates in the operational environment. This can be done by using a dedicated software, e.g. CREME- 96 [10], by assigning the environment and entering the parameterized SEU cross-section curve. From these input data, the code will give an estimation for the SEU rate in orbit. A very good demonstration of this procedure is given in Ref. [104], where one of the studied devices is the 4 Mbit Atmel AT60142F SRAM, discussed above. The agreement reported in this chapter between estimated and observed SEU rates is fairly good.

1.5.1 Beam Requirements

As mentioned in the previous sections, the radiation source used for a particular radiation test must fulfil a certain set of requirements, which depends

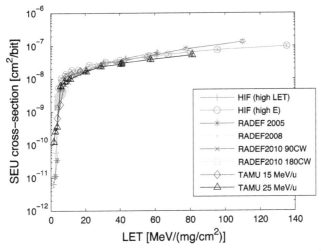

Figure 1.21   SEU cross section for 4 Mbit Atmel AT60142F SRAM used in the SEU monitor [52]. Figure taken from Ref. [4].

on the type of test to be carried out. The following subsections give a brief summary of the ESA ECSS requirements for parts qualification in TID, TNID and SEE tests. The dose level or fluence level required to qualify a part is generally dependent on its intended mission.

1.5.2 TID Tests

TID tests are generally carried out with electrons or gamma rays; the ESA ESCC Basic Specification No. 22900 [105] indicates that TID tests may be carried out with either cobalt-60 gamma rays (two peak photon energies at 1.17 MeV and 1.33 MeV, with a low-energy tail) or with steady-state electron irradiation at an energy of at least 1 MeV at the sensitive volume. At these energies, photons and electrons have the advantage of high penetration (the typical component thickness is much smaller than the attenuation length/path length) and do not cause considerable displacement damage.

The dose rate can be adjusted to meet the required dose level in a reasonable irradiation time – ESCC specifications allow a range of 36 rad/h to 180 krad/h. However, the selected dose rate must take into account an eventual enhanced low dose rate sensitivity (ELDRS) [106] of the tested device. The irradiation must be carried out at room temperature and can be carried out in air.

1.5.3 TNID Tests

Currently, no standard method exists for TNID testing. However, ESA standards ECSS-Q-ST-60-15C and ECSS-E-HB-10-12A recommend that TNID tests be carried out with protons, at several different energies, up to 200 MeV, for a total TNID level of $2 \cdot 10^{11}$ cm^{-2} 50 MeV equivalent proton fluence. The equivalent 50 MeV proton fluence can be calculated with the methods described in [107–108].

1.5.4 SEE Tests

Single Event Effects tests are carried out using protons and heavy ions. According to ESCC Basic Specification No. 25100:

- Proton SEE testing must be carried out using several different energies, ranging from a few MeV (to check sensitivity to direct proton ionization) to 200 MeV. Ideally, the source should be able to deliver a variable particle flux in the range of 10^5 cm$^{-2} \cdot s^{-1}$ to 10^8 cm$^{-2} \cdot s^{-1}$. It is recommended to check the sensitivity of the DUT to the beam incidence angle.
- Heavy-ion testing must be carried out using ions with a range of at least 40 µm in silicon, with a variable flux ranging between a few tens of ions.cm$^{-2} \cdot s^{-1}$ to 10^5 ions $\cdot$ cm$^{-2} \cdot s^{-1}$. Ideally, the irradiation should be performed in vacuum to avoid degrading the beam spectrum, and the ion energy should be selected so that the Bragg peak is located at the depth of the device's sensitive layers.

1.5.5 Sample Preparation

Depending on the radiation type and energy to be used during testing, sample preparation steps may be necessary to allow an adequate energy deposition profile within the device.

1.5.6 TID Tests

In thin target devices, the secondary electrons generated by the primary electron beam may have sufficient energy to escape the target material, resulting in a loss of energy from the target and in a reduction of the effective deposited dose. To counter this effect, it is recommended to surround target devices with material providing charged-particle equilibrium during irradiation.

In the case of cobalt-60 irradiation, the upper layers of the device may receive a higher dose because of the low-penetration, low-energy tail of the cobalt-60 gamma spectrum. To mitigate this effect, it is advisable to surround the target device with a container of at least 1.5 mm of lead, with at least 0.7 mm of aluminium lining [105].

1.5.7 SEE Tests

SEE tests carried out with low-penetration beams (e.g. heavy-ion and low-energy proton beams) must be carried out on "opened" packages, to minimize the amount of material in the beam's path – thereby minimizing the energy spread of the beam. Package opening is generally performed via a combination of mechanical and laser ablation, and chemical etching. It is a delicate operation, because the device must remain functional after opening – meaning that the delicate chip features must not be damaged, in particular the thin bonding wires connecting the die to the package pins.

SEE tests carried out with high-penetration beams (e.g. neutron and high-energy proton beams) may be carried out on non-opened devices.

1.5.8 Radiation Facilities

Introducing different European facilities for radiation effects studies as of 2018. Discussing in detail the RADEF facility of University of Jyväskylä, Finland.

1.5.9 ESA European Component Irradiation Facilities (ECIF)

The European Space Agency supports four European Component Irradiation Facilities (ECIF) to perform its radiation hardness testing:

- The Radiation Effects Facility (RADEF), at the University of Jyväskylä(Finland), can provide a variety of particle beams [109]:
  - heavy-ion beams from nitrogen to xenon, at energies from 4 to 9.3 MeV/amu (for a surface LET in silicon ranging from 1.83 to 60.0 MeV $\cdot$ cm^{-2} $\cdot$ mg^{-1});
  - proton beams from about 500 keV to 50 MeV;
  - electron beams at 6, 9, 12, 16 and 20 MeV;
  - continuous spectrum X-ray beams – either from 0 to 6 MeV, with a peak around 1 MeV, or from 0 to 15 MeV, with a peak around 2 MeV.

- Several facilities at the Université Catholique de Louvain (UCL, Belgium):

    o the Heavy Ion Facility (HIF), which provides ions from carbon to xenon (for a surface LET in silicon ranging from 1.3 to 62.5 MeV $\cdot$ cm^{-2} $\cdot$ mg^{-1});
    o the Light Ion Facility (LIF), which provides a 65 MeV proton beam, which can be degraded down to about 10 MeV;
    o the Neutron Irradiation Facility, which provides continuous-spectrum secondary neutron fields;
    o a cobalt-60 source for TID testing.

- The Proton Irradiation facility (PIF, Paul Scherrer Institute, Switzerland) [110], which delivers proton beams from 6 to 230 MeV
- The cobalt-60 irradiation facility at the ESA European Space Research and Technology Centre (ESTEC), which can be used for TID testing.

1.5.10 Other Outstanding European Facilities

Among the main European accelerator facilities used for space radiation physics are:

- The UNILAC linear accelerator and SIS-18 synchrotron at the Gesellschaft für Schwerionenforschung (GSI, Darmstadt, Germany), which can deliver a wide variety of heavy-ion beams ranging from 1 MeV/amu to 1 GeV/amu. The low-energy ions from the UNILAC can be collimated down to a 10 μm diameter spot for spatially resolved radiation hardness studies.
- Several facilities at the Grand Accélérateur National d'Ions Lourds (GANIL, Caen, France), which together can accelerate heavy ions from carbon to lead at energies ranging from below 1 keV/amu up to 100 MeV/amu, generating LETs in silicon from below 1 up to 100 MeV $\cdot$ cm^2 $\cdot$ mg^{-1}.
- The U-400M cyclotron of the Joint Institute for Nuclear Research (JINR, Dubna, Russia) can accelerate ions from lithium to bismuth at energies ranging from 3 to 40 MeV/amu, generating LETs in silicon from about 1 to 100 MeV $\cdot$ cm^2 $\cdot$ mg^{-1}.
- The ChipIR facility at the ISIS Neutron and Muon Source (Rutherford Appleton Laboratory, Didcot, UK) is a beamline dedicated to the study of SEEs in microelectronics, using an atmospheric-spectrum neutron beam.

1.5.11 Other Outstanding Facilities in the World

- The 20-ID-B2 beamline at the Advanced Photon Source (APS, Argonne National Laboratory, Chicago, USA-IL) and the VESPERS beamline at the Canadian Light Source (CLS, Saskatoon, Canada) can deliver pulsed, focused X-ray beams for SEE testing. This is similar to laser SEE testing, with the added benefit that unlike laser light, X-rays can penetrate the metal layers covering the sensitive volumes in highly integrated ICs.
- The BASE facility at the 88-inch Cyclotron (Lawrence Berkeley National Laboratory, Berkeley, USA-CA) is specialized in radiation effects testing. It can provide neutron beams from 8 to 30 MeV, proton beams from a few MeV to 55 MeV, and heavy ions from helium to bismuth, at energies between 4.5 and 30 MeV/amu, generating LETs in silicon between 0.1 and 100 MeV $\cdot$ cm^2 $\cdot$ cm^{-1}.
- The Crocker Nuclear Laboratory (University of California Davis, Davis, USA-CA) has been studying the effects of space radiation on electronics since the 1970s; its 76-inch cyclotron can provide proton beams from 4 to 67.5 MeV, in addition to neutron and alpha beams.
- The Radiation Effects Facility at the Texas A&M University (TAMU, College Station, USA-TX) was designed to irradiate electronics components with ions ranging from helium to gold at energies between up to 40 MeV/amu, yielding surface LETs in silicon between 0.11 and 82 MeV $\cdot$ cm^2 $\cdot$ mg^{-1}. It can also deliver proton beams up to 40 MeV.
- The Proton Irradiation Facility (PIF) and Neutron Irradiation Facility (NIF) at TRIUMF (Vancouver, Canada), which can be used for high-energy proton and atmospheric-spectrum neutron SEE studies.
- The Single Event Effects Test Facility (SEETF) at the Michigan State University, which can be used for heavy-ion SEE testing [111].
- The Single Event Effects test facility at Oak Ridge National Laboratory, which can be used for high-energy proton testing of large equipment [112].

1.5.12 Accelerated Test for Memories

In the following part of this chapter, experimental results from radiation testing campaigns using different types of particle accelerators will be given and analyzed. Before exposing these results, it is important to describe some details on experiment configuration during the test of memory devices.

1.5.13 Memory Test Setup

The accelerated test on memories can be carried out to explore the effects of TID or the occurrence of single events. The TID tests do not require the monitoring of the device during the irradiation phase, but rather measures or functional tests that can be periodical operated when the device is not irradiated. The test of memories for single events, especially in dynamic mode, requires a continuous monitoring. For this purpose, a test setup similar to the one depicted in Figure 1.22 must be used. In this setup, the chip to be irradiated is mounted on a dedicated PCB card and controlled by a tester board via cables that are long enough to keep the tester outside of the beamline. Such a setup is chosen to assure the correct functionality of the tester during irradiation. The Tester board has a Field Programmable Gate Array (FPGA) as the main processing and communication unit, which the test firmware is implemented. FPGAs are chosen as testers more often than microcontrollers, thanks to the reliability of their operation flow, and the possibility of complete control over the timing of execution of operations, among other benefits. Both the tester board and the memory board are powered with power supplies that are located in the control room (or at least driven from it), in order to be able to power them off in case of a persisting SEL.

When it comes to static mode testing, the test firmware writes the data background sequence to the memory device prior to irradiation. After the beam exposure, it reads the data back, and compares them to the initial data

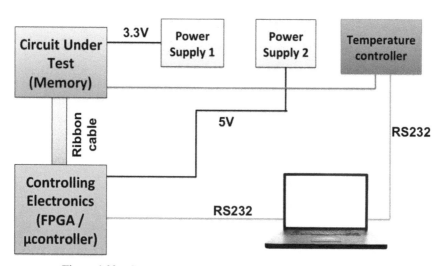

Figure 1.22   General scheme of irradiation test setup for memories.

pattern. In case of an error, the tester sends to the control computer all the necessary information for data processing. More specifically, the message is composed of an initiating sequence that indicates the code of the device, the address of the erroneous word, the erroneous word and finally information regarding the applied test. In the case of static mode test, the expected data background is sent, while in the case of dynamic mode test, made through March algorithms, the element and operation at which the error was located is given (e.g. the third element and the first operation of the March C- algorithm, "r1", see Table 1.2), as well as the timestamp of the error occurrence.

One of the essential parameters for the data processing of the observed SEUs is the knowledge of the scrambling algorithm of the memory. Without the knowledge of the actual physical location of the SRAM cells in the memory array, it is very difficult not only to distinguish between SBUs and MCUs, but also to understand the origins of larger events occurring during dynamic mode testing such as SELs and SEFIs.

Adapting March algorithms for the purpose of radiation testing is a straightforward process. The tester needs to implement a Finite State Machine (FSM) which, depending on the algorithm, applies the required operations and elements to the memory under test. Besides the stimulus action, the read operations of the March algorithm work also as verification elements of the test, since the read pattern is compared with the expected one. For example, for a March element like ↓(r0, w1), the read "0" operation will read the contents of an SRAM word, and will compare them to a "golden reference" word with all its bits being "0"s. In case of an upset, a message described earlier will be transmitted to the control computer. For both the static and dynamic mode tests, it is imperative to record all addresses and contents of the failing words, and not to use an internal counter when testing SRAMs for the aforementioned reasons of detecting and identifying MCUs, SEFIs, SELs and other large-scale events.

Although stimulating the memory by using various testing techniques is one of the most important parameters to be considered when performing radiation testing, it is not the only factor that can affect the sensitivity of the device. Temperature variations can have an effect, either increasing or decreasing their performance. In order to achieve temperature variations when placing the memory devices under the beam, a complementary part of the setup is the temperature controller. A feedback loop composed of a foil heater, a thermocouple as a temperature sensor and a controlling instrument can be used to achieve high temperature variations. The sensor would provide the current temperature to the controlling instrument, which would increase or decrease the input current to the foil heater to achieve the target temperature. Low temperature variations can be achieved by inserting the memory in a cryogenic chamber filled with argon gas that allows decreasing the temperature below 0°C, preventing the condensation of water vapour on the electronic devices. With a foil heater, a temperature sensor and a control instrument, the desired temperature is achieved.

Depending on the energy of the particles used to test the memory, the die of the device may need to be directly exposed, to avoid excessive energy straggling or even complete attenuation of the beam before reaching the silicon. When it comes to heavy ions or low-energy proton testing, the DUT is inserted into a vacuum chamber, and it is often necessary that the top part of the packaging is removed (an operation referred to as "delidding"). In order to be able to change the device in a relatively easy manner when the dose levels are high, a chip socket is used to hold the chip.

1.5.14 Notes on Test Result Analysis

Regarding the irradiation of semiconductor memories, the primary result that is generally analyzed is the cross-section of the device, as was introduced earlier in this chapter. The calculation of the cross-section is done using the number of upsets observed in the memory and the total particle fluence per unit area. Nevertheless, it is important to note that a single-event may either lead to one Single-Bit Upset, or to several bit flips (Multiple Cell Upset). Thus, it is important to distinguish a **raw-cross section**, which considers each bit flip in the computation, and an **event-cross section,** which considers only the event count (each event causing one or more bit flips). This concept is well understandable when the test bitmap of the irradiated memory is analyzed. An example is given by the bitmap depicted in Figure 1.23, which was generated from the results of heavy-ion dynamic test irradiation on a 90 nm, 32 MB SRAM from Cypress Semiconductors. Several small patches and a few large bands are visible, where the data have been severely corrupted. A detailed analysis of these different types of error clusters is available in [113], which will be briefly reported here. The smallest patches, barely visible, have been caused by direct heavy-ion ionization (type A). The small horizontal patches were caused by limited micro-latch-ups (type B). The wide horizontal bands were very likely caused by a failure of the I/O data buffers (type C), and the wide vertical bands on the edges of the die were probably caused by a failure of the power switches (type D).

Provided that the memory has not suffered too many upsets during a test (the physical bitmap does not exhibit too many sparse black pixels),

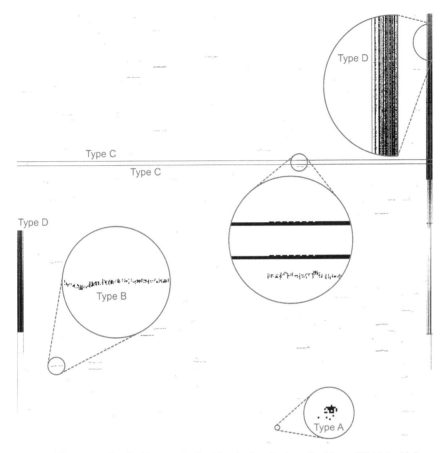

Figure 1.23 Example of a bitmap obtained by the irradiation of a 90 nm SRAM with heavy ions. Each pixel corresponds to a bit in the memory array, and every black pixel represents a corrupted bit. The image is 4096 × 4096 pixels.

it is possible to systematically group together neighbouring upsets in clusters. Counting the clusters provides an accurate estimate of the number of single events that occurred during irradiation.

1.6 Conclusion

In this chapter, we introduced some elements to explain the importance of the characterization of radiation effects on electronic devices, and in particular on memories. Moreover, basic concepts such as single-event effects, total ionizing dose effects and fault mechanisms have been introduced, in order to ease the understanding of the following chapters.

References

[1] ACE Science Team, "ACE Level 2 Data."

[2] D. H. Hathaway, "NASA/Marshall Solar Physics."

[3] E. C. Stone, A. M. Frandsen, R. A. Mewaldt, E. R. Christian, D. Margolies, J. F. Ormes, and F. Snow, "The Advanced Composition Explorer," Space Sci. Rev., vol. 86, no. 1/4, pp. 1–22, 1998.

[4] A. Javanainen, "Particle radiation in microelectronics," University of Jyväskylä, 2012.

[5] J. R. Jokipii, C. P. Sonett, and M. S. Giampapa, Cosmic winds and the heliosphere. University of Arizona Press, 1997.

[6] E. C. Stone, C. M. S. Cohen, W. R. Cook, A. C. Cummings, B. Gauld, B. Kecman, R. A. Leske, R. A. Mewaldt, M. R. Thayer, B. L. Dougherty, R. L. Grumm, B. D. Milliken, R. G. Radocinski, M. E. Wiedenbeck, E. R. Christian, S. Shuman, and T. T. von Rosenvinge, "The Solar Isotope Spectrometer for the Advanced Composition Explorer," Space Sci. Rev., vol. 86, no. 1/4, pp. 357–408, 1998.

[7] J. A. Van Allen, C. E. McIlwain, and G. H. Ludwig, "Radiation observations with satellite 1958?," J. Geophys. Res., vol. 64, no. 3, pp. 271–286, Mar. 1959.

[8] British Geological Survey - Natural Environment Research Council, "Magnetic Poles."

[9] SPENVIS Collaboration, "The Space Environment Information System".

[10] The CREME Collaboration, "CREME-MC site."

[11] A. Viljanen and R. Pirjola, "Geomagnetically induced currents in the Finnish high-voltage power system," Surv. Geophys., vol. 15, no. 4, pp. 383–408, Jul. 1994.

[12] V. F. Hess, "Über Beobachtungen der durchdringenden Strahlung bei sieben Freiballonfahrten," Phys. Z., vol. 13, pp. 1084–1091, 1912.

[13] J. F. Ziegler and W. A. Lanford, "Effect of Cosmic Rays on Computer Memories," Science (80-.)., vol. 206, no. 4420, pp. 776–788, Nov. 1979.

[14] E. Normand, "Single event upset at ground level," IEEE Trans. Nucl. Sci., vol. 43, no. 6, pp. 2742–2750, 1996.

[15] T. C. May and M. H. Woods, "Alpha-particle-induced soft errors in dynamic memories," IEEE Trans. Electron Devices, vol. 26, no. 1, pp. 2–9, Jan. 1979.

[16] J. F. Ziegler, H. W. Curtis, H. P. Muhlfeld, C. J. Montrose, B. Chin, M. Nicewicz, C. A. Russell, W. Y. Wang, L. B. Freeman, P. Hosier, L. E. LaFave, J. L. Walsh, J. M. Orro, G. J. Unger, J. M. Ross, T. J. O'Gorman, B. Messina, T. D. Sullivan, A. J. Sykes, H. Yourke, T. A. Enger, V. Tolat, T. S. Scott, A. H. Taber, R. J. Sussman, W. A. Klein, and C. W. Wahaus, "IBM experiments in soft fails in computer electronics (1978–1994)," IBM J. Res. Dev., vol. 40, no. 1, pp. 3–18, Jan. 1996.

[17] E. Ibe, H. Taniguchi, Y. Yahagi, K. Shimbo, and T. Toba, "Impact of Scaling on Neutron-Induced Soft Error in SRAMs From a 250 nm to a 22 nm Design Rule," IEEE Trans. Electron Devices, vol. 57, no. 7, pp. 1527–1538, Jul. 2010.

[18] M. Mason, "Cosmic rays damage automotive electronics." May-2006.

[19] L. Dominik, E. Normand, M. J. Dion, and P. Ferguson, "Proposal for a new integrated circuit and electronics neutron experiment source at oak ridge national laboratory," in 2009 IEEE International Reliability Physics Symposium, 2009, pp. 940–947.

[20] Bloomberg, "Delta, Qantas, Air Canada Divert Flights Due to Solar Storm," Jan-2012.

[21] E. Normand, "Single-event effects in avionics," IEEE Trans. Nucl. Sci., vol. 43, no. 2, pp. 461–474, Apr. 1996.

[22] "The IEE Seminar on Cosmic Radiation Single Event Effects and Avionics, 2005 (Ref. No. 2005/11270)," 2005, p. 0-1.

[23] P. K. F. Grieder, Cosmic rays at Earth: researcher's reference manual and data book. Elsevier Science Ltd., 2001.

[24] B. D. Sierawski, M. H. Mendenhall, R. A. Reed, M. A. Clemens, R. A. Weller, R. D. Schrimpf, E. W. Blackmore, M. Trinczek, B. Hitti, J. A. Pellish, R. C. Baumann, S.-J. Wen, R. Wong, and N. Tam, "Muon-Induced Single Event Upsets in Deep-Submicron Technology," IEEE Trans. Nucl. Sci., vol. 57, no. 6, pp. 3273–3278, Dec. 2010.

[25] A. Hillier, K. Ishida, A. Bosser, and L. Dilillo, "Improving the Stability of Electronic Systems – Measuring the Effects of Muons on Static Random Access Memory," *Proceedings of the 14th International Conference on Muon Spin Rotation, Relaxation and Resonance*, Sapporo, Japan, 2017.

[26] A. Bosser, "Single-Event Effects of Space and Atmospheric Radiation on Memory Components", Doctoral dissertation, University of Jyväskylä/University of Montpellier, 2017.

[27] J. C. Kotz, P. Treichel, and J. R. Townsend, Chemistry and chemical reactivity, 7th ed. Thomson Brooks/Cole, 2009.

[28] G. A. Sai-Halasz, M. R. Wordeman, and R. H. Dennard, "Alpha-particle-induced soft error rate in VLSI circuits," IEEE Trans. Electron Devices, vol. 29, no. 4, pp. 725–731, Apr. 1982.

[29] K. Itoh, R. Hori, H. Masuda, Y. Kamigaki, H. Kawamoto, and H. Katto, "A single 5V 64K dynamic RAM," in 1980 IEEE International Solid-State Circuits Conference. Digest of Technical Papers, 1980, vol. XXIII, pp. 228–229.

[30] S. Martinie, J. L. Autran, S. Uznanski, P. Roche, G. Gasiot, D. Munteanu, and S. Sauze, "Alpha-Particle Induced Soft-Error Rate in CMOS 130 nm SRAM," IEEE Trans. Nucl. Sci., vol. 58, no. 3, pp. 1086–1092, Jun. 2011.

[31] R. Wong, P. Su, S.-J. Wen, B. Dwyer McNally, and S. Coleman, "The effect of Radon on soft error rates for wire bonded memories," in 2010 IEEE International Integrated Reliability Workshop Final Report, 2010, pp. 133–134.

[32] I. Batkin, R. B. del Re, J.-G. Boutin, and J. Armitage, "Gamma-spectroscopy investigation of radon daughter deposition on electrostatically charged surfaces," Phys. Med. Biol., vol. 43, no. 3, pp. 487–499, Mar. 1998.

[33] E. Segrè, "Fermi and Neutron Physics," Rev. Mod. Phys., vol. 27, no. 3, pp. 257–263, 1955.

[34] R. C. Baumann and E. B. Smith, "Neutron-induced 10B fission as a major source of soft errors in high density SRAMs," Microelectron. Reliab., vol. 41, no. 2, pp. 211–218, Feb. 2001.

[35] L. Berkhouse, S. E. Davis, F. R. Gladeck, J. H. Hallowell, and C. B. Jones, "Operation DOMINIC I-1962," Feb. 1983.

[36] S. L. Simon and W. L. Robison, "A compilation of nuclear weapons test detonation data for US Pacific Ocean tests," Heal. Phys., vol. 73, no. 1, pp. 258–264, 1997.

[37] P. Sigmund, Stopping of heavy ions : a theoretical approach. Springer, 2004.

[38] P. Sigmund, "Stopping Power: Wrong Terminology," ICRU News, 2000.

[39] P. Sigmund, Particle Penetration and Radiation Effects – General Aspects and Stopping of Swift Point Charges. Springer, 2006.

[40] N. Bohr, "On the theory of the decrease of velocity of moving electrified particles on passing through matter," Philos. Mag. Ser. 6, vol. 25, no. 145, pp. 10–31, 1913.

[41] N. Bohr, "On the decrease of velocity of swiftly moving electrified particles in passing through matter," Philos. Mag. Ser. 6, vol. 30, no. 178, pp. 581–612, 1915.

[42] N. Bohr, "The penetration of atomic particles through matter," Kgl. Danske Vidensk. Selsk. Mat.-Fys. Medd., vol. 18, no. 8, pp. 1–144, 1948.

[43] J. Lindhard and A. H. Sorensen, "Relativistic theory of stopping for heavy ions," Phys. Rev. A, vol. 53, no. 4, pp. 2443–2456, 1996.

[44] F. Yang and J. H. Hamilton, Modern Atomic and Nuclear Physics. World Scientific Pub Co., 2009.

[45] J. Lilley, Nuclear Physics - Principles and Applications. John Wiley & Sons Ltd., 2001.

[46] P. E. Dodd, J. R. Schwank, M. R. Shaneyfelt, V. Ferlet-Cavrois, P. Paillet, J. Baggio, G. L. Hash, J. A. Felix, K. Hirose, and H. Saito, "Heavy Ion Energy Effects in CMOS SRAMs," IEEE Trans. Nucl. Sci., vol. 54, no. 4, pp. 889–893, Aug. 2007.

[47] Y. P. Varshni, "Temperature dependence of the energy gap in semiconductors," Physica, vol. 34, no. 1, pp. 149–154, Jan. 1967.

[48] T. H. DiStefano and D. E. Eastman, "The band edge of amorphous SiO_2 by photoinjection and photoconductivity measurements," Solid State Commun., vol. 9, no. 24, pp. 2259–2261, Dec. 1971.

[49] R. C. Alig and S. Bloom, "Electron-Hole-Pair Creation Energies in Semiconductors," Phys. Rev. Lett., vol. 35, no. 22, pp. 1522–1525, Dec. 1975.

[50] J. M. Benedetto and H. E. Boesch, "The Relationship between 60Co and 10-keV X-Ray Damage in MOS Devices," IEEE Trans. Nucl. Sci., vol. 33, no. 6, pp. 1317–1323, 1986.

[51] R. Harboe-Sorensen, C. Poivey, F.-X. Guerre, A. Roseng, F. Lochon, G. Berger, W. Hajdas, A. Virtanen, H. Kettunen, and S. Duzellier, "From the Reference SEU Monitor to the Technology Demonstration Module On-Board PROBA-II," IEEE Trans. Nucl. Sci., vol. 55, no. 6, pp. 3082–3087, Dec. 2008.

[52] R. Harboe-Sorensen, F.-X. Guerre, and A. Roseng, "Design, Testing and Calibration of a 'Reference SEU Monitor' System," in 2005 8th

European Conference on Radiation and Its Effects on Components and Systems, 2005, pp. B3-1-B3-7.

[53] G. C. Messenger and M. S. Ash, Single event phenomena. Chapman & Hall, 1997.

[54] M. A. Xapsos, "A Spatially Restricted Linear Energy Transfer Equation," Radiat. Res., vol. 132, no. 3, p. 282, Dec. 1992.

[55] M. A. Xapsos, "Applicability of LET to single events in microelectronic structures," IEEE Trans. Nucl. Sci., vol. 39, no. 6, pp. 1613–1621, 1992.

[56] M. Xapsos, G. Summers, E. Burke, and C. Poivey, "Microdosimetry theory for microelectronics applications," Nucl. Instruments Methods Phys. Res. Sect. B Beam Interact. with Mater. Atoms, vol. 184, no. 1–2, pp. 113–134, Sep. 2001.

[57] J. J. Butts and R. Katz, "Theory of RBE for Heavy Ion Bombardment of Dry Enzymes and Viruses," Radiat. Res., vol. 30, no. 4, pp. 855–871, Apr. 1967.

[58] R. Katz, "Track structure theory in radiobiology and in radiation detection," Nucl. Track Detect., vol. 2, no. 1, pp. 1–28, Mar. 1978.

[59] F. A. Cucinotta, R. Katz, and J. W. Wilson, "Radial distribution of electron spectra from high-energy ions," Radiat. Environ. Biophys., vol. 37, no. 4, pp. 259–265, Dec. 1998.

[60] B. Grosswendt, S. Pszona, and A. Bantsar, "New descriptors of radiation quality based on nanodosimetry, a first approach," Radiat. Prot. Dosimetry, vol. 126, no. 1–4, pp. 432–444, May 2007.

[61] K. M. Prise, G. Schettino, M. Folkard, and K. D. Held, "New insights on cell death from radiation exposure," Lancet Oncol., vol. 6, no. 7, pp. 520–528, Jul. 2005.

[62] "The International Technology Roadmap for Semiconductors."

[63] Intel, "Microprocessor Quick Reference Guide."

[64] J. F. Ziegler, J. P. Biersack, M. Ziegler, D. J. Marwick, G. A. Cuomo, W. A. Porter, and S. A. Harrison, "{SRIM} 2013 code," vol. SRIM-2013.

[65] V. Zajic, "TVDG LET Calculator," vol. 1.24. 2001.

[66] V. Zajic and P. Thieberger, "Heavy ion linear energy transfer measurements during single event upset testing of electronic devices," IEEE Trans. Nucl. Sci., vol. 46, no. 1, pp. 59–69, Feb. 1999.

[67] A. Javanainen, W.~H.~Trzaska, R.~Harboe-Sørensen, A. Virtanen, G.~Berger, and W.~Hajdas, "ECIF Cocktail Calculator."

[68] A. Javanainen, "A simple expression for electronic stopping force of heavy ions in solids," Nucl. Instruments Methods Phys. Res. Sect. B Beam Interact. with Mater. Atoms, vol. 285, pp. 158–161, Aug. 2012.

[69] P. Sigmund, "{PASS} code."

[70] P. Sigmund and A. Schinner, "Binary theory of electronic stopping," Nucl. Instruments Methods Phys. Res. Sect. B Beam Interact. with Mater. Atoms, vol. 195, no. 1–2, pp. 64–90, Oct. 2002.

[71] G. Schiwietz and P. L. Grande, "Convolution approximation for swift Particles."

[72] P. Helmut, "Stopping Power of Matter for Ions."

[73] F. Hubert, R. Bimbot, and H. Gauvin, "Range and stopping-power tables for 2.5–500 MeV/nucleon heavy ions in solids," At. Data Nucl. Data Tables, vol. 46, no. 1, pp. 1–213, Sep. 1990.

[74] Geant4 Collaboration, "Geant4: A toolkit for the simulation of the passage of particles through matter."

[75] A. Javanainen, T. Malkiewicz, J. Perkowski, W. H. Trzaska, A. Virtanen, G. Berger, W. Hajdas, R. Harboe-Sørensen, H. Kettunen, V. Lyapin, M. Mutterer, A. Pirojenko, I. Riihimàki, T. Sajavaara, G. Tyurin, and H. J. Whitlow, "Linear energy transfer of heavy ions in silicon," IEEE Trans. Nucl. Sci., vol. 54, no. 4, pp. 1158–1162, Aug. 2007.

[76] T. R. Oldham and F. B. McLean, "Total ionizing dose effects in MOS oxides and devices," IEEE Trans. Nucl. Sci., vol. 50, no. 3, pp. 483–499, Jun. 2003.

[77] R. L. Pease, "Total ionizing dose effects in bipolar devices and circuits," IEEE Trans. Nucl. Sci., vol. 50, no. 3, pp. 539–551, Jun. 2003.

[78] "Test Procedures for the Measurement of Single-Event Effects in Semiconductor Devices from Heavy Ion Irradiation." 1996.

[79] "Measurement and Reporting of Alpha Particle and Terrestrial Cosmic Ray-Induced Soft Errors in Semiconductor Devices." 2006.

[80] P. C. Adell, R. D. Schrimpf, C. R. Cirba, W. T. Holman, X. Zhu, H. J. Barnaby, and O. Mion, "Single event transient effects in a voltage reference," Microelectron. Reliab., vol. 45, no. 2, pp. 355–359, Feb. 2005.

[81] N. D. P. Avirneni and A. Somani, "Low Overhead Soft Error Mitigation Techniques for High-Performance and Aggressive Designs," IEEE Trans. Comput., vol. 61, no. 4, pp. 488–501, Apr. 2012.

[82] M. Nicolaidis, Soft Errors in Modern Electronic Systems. Springer, 2010.

[83] F. W. Sexton, "Destructive single-event effects in semiconductor devices and ICs," IEEE Trans. Nucl. Sci., vol. 50, no. 3, pp. 603–621, Jun. 2003.

[84] J. R. Schwank, V. Ferlet-Cavrois, M. R. Shaneyfelt, P. Paillet, and P. E. Dodd, "Radiation effects in SOI technologies," IEEE Trans. Nucl. Sci., vol. 50, no. 3, pp. 522–538, Jun. 2003.

[85] A. Bosio, L. Dilillo, P. Girard, S. Pravossoudovitch, and A. Virazel, "Advanced Test Methods for SRAMs – Effective Solutions for Dynamic Fault Detection in Nanoscaled Technologies." Springer, 2010.

[86] F. Wrobel, J.-M. Palau, M. "Induced nuclear reactions in a simplified SRAM structure: scaling effects on SEU and MBU" IEEE Trans. Nucl. Sci., vol. 48, no. 6, pp. 1946–1952, Dec. 2001.

[87] R. C. Baumann "radiation induced soft errors in advanced semiconductor technologies", IEEE Trans. Device Mater. Reliab., vol. 5, no. 3, pp. 305–316, Sep. 2005.

[88] T. Heijmen et al. "Factors that impact the critical charge of memory elements", in proc. IOLTS 2006, pp. 57–62.

[89] "Space Product Assurance - Commercial electrical, electronic and electromechanical (EEE) components." ECSS-Q-ST-60-13C, 21-Oct-2013.] [JESD89, "Measurement and reporting of alpha particles and terrestrial cosmic ray induced Soft Error in Semiconductor devices." JEDEC Standard.

[90] "Test Procedures for the Measurement of Single-Event Effects in Semiconductor Devices from Heavy Ion Irradiation." JEDEC Test Standard, No 57, Dec-1996.

[91] D. Radaelli, H. Puchner, S. Wong, and S. Daniel, Investigation of multi-bit upsets in a 150 nm technology SRAM device, IEEE Trans. Nucl. Sci., vol. 52, no. 6, pp. 2433–2437, Dec. 2005.

[92] P. Rech, J.-M. Galliere, P. Girard, A. Griffoni, J. Boch, F. Wrobel, F. Saigne, and L. Dilillo, "Neutron-Induced Multiple Bit Upsets on Two Commercial SRAMs Under Dynamic-Stress," IEEE Trans. Nucl. Sci., vol. 59, no. 4, pp. 893–899, Aug. 2012.

[93] V. Gupta, A. Bosser, G. Tsiligiannis, A. Zadeh, A. Javanainen, A. Virtanen, H. Puchner, F. Saigné, F. Wrobel and L. Dilillo, "Heavy-ion radiation impact on a 4Mb FRAM under different test conditions",

European Conference on Radiation and its Effects on Components and Systems (RADECS), Moscow, Russia, Sept. 2015.

[94] L. Dilillo, G. Tsiligiannis, V. Gupta, A. Bosser, F. Saigné and F. Wrobel, "Soft errors in commercial off-the-shelf static random access memories", Journal of Semiconductor Science and Technology, vol 32, I 1, DOI: 10.1088/1361-6641/32/1/013006, 2016.

[95] "Single Event Effects Test Method and Guidelines," ESCC Basic Specif. No. 25100, 2002.

[96] "Total Dose Steady-State Irradiation Test Method." Oct-2010.

[97] "Test Method Standard-Microcircuits." 2010.

[98] "Standard Guide for the Measurement of Single Event Phenomena (SEP) Induced by Heavy Ion Irradiation of Semiconductor Devices." 2006.

[99] A. J. Tylka, J. H. Adams, P. R. Boberg, B. Brownstein, W. F. Dietrich, E. O. Flueckiger, E. L. Petersen, M. A. Shea, D. F. Smart, and E. C. Smith, "CREME96: a revision of the Cosmic Ray Effects on Micro-Electronics code," IEEE Trans. Nucl. Sci., vol. 44, no. 6, pp. 2150–2160, 1997.

[100] International Atomic Energy Agency – Nuclear Data Service, "Live Chart of Nuclides".

[101] G. Berger, "Heavy Ion Irradiation Facility (HIF)." 2004.

[102] A. Virtanen, H. Kettunen, A. Javanainen, M. Rossi, and J. Jaatinen, "RADiation Effects Facility at JYFL".

[103] "Radiation Effects Facility/The Cyclotron Institute/Texas A&M University".

[104] R. Harboe-Sorensen, C. Poivey, A. Zadeh, A. Keating, N. Fleurinck, K. Puimege, F.-X. Guerre, F. Lochon, M. Kaddour, L. Li, and D. Walter, "PROBA-II Technology Demonstration Module In-Flight Data Analysis," IEEE Trans. Nucl. Sci., vol. 59, no. 4, pp. 1086–1091, Aug. 2012.

[105] European Space Components Coordination, "Total Dose Steady-State Irradiation Test Method – ESCC Basic Specification No. 22900," 2010.

[106] D. M. Fleetwood, S. L. Kosier, R. N. Nowlin, R. D. Schrimpf, R. A. Reber, M. DeLaus, P. S. Winokur, A. Wei, W. E. Combs, and R. L. Pease, "Physical mechanisms contributing to enhanced bipolar gain degradation at low dose rates," IEEE Trans. Nucl. Sci., vol. 41, no. 6, pp. 1871–1883, Dec. 1994.

[107] I. Jun, M. A. Xapsos, S. R. Messenger, E. A. Burke, R. J. Walters, G. P. Summers, and T. Jordan, "Proton Nonionizing Energy Loss (NIEL) for

Device Applications," in IEEE Transactions on Nuclear Science, 2003, vol. 50, no. 6 I, pp. 1924–1928.

[108] S. R. Messenger, E. A. Burke, M. A. Xapsos, G. P. Summers, R. J. Walters, I. Jun, and T. Jordan, "NIEL for Heavy Ions: An Analytical Approach," in IEEE Transactions on Nuclear Science, 2003, vol. 50, no. 6 I, pp. 1919–1923.

[109] University of Jyväskylä, "RADEF – Radiation Effects Facility." [Online]. Available: https://www.jyu.fi/science/en/physics/research/infr astructures/accelerator-laboratory/radiation-effects-facility. [Accessed: 13-Mar-2018].

[110] Paul Scherrer Institute, "PIF webpage." [Online]. Available: https://www.psi.ch/pif/. [Accessed: 13-Mar-2018].

[111] R. Ladbury, R. A. Reed, P. Marshall, K. A. LaBel, R. Anantaraman, R. Fox, D. P. Sanderson, A. Stolz, J. Yurkon, A. F. Zeller, and J. W. Stetson, "Performance of the high-energy single-event effects test Facility (SEETF) at Michigan State university's national Super-conducting Cyclotron laboratory (NSCL)," IEEE Trans. Nucl. Sci., vol. 51, no. 6, pp. 3664–3668, Dec. 2004.

[112] B. W. Riemer, F. X. Gallmeier, and L. J. Dominik, "Single Event Effects test facility at Oak Ridge National Laboratory," in 2015 IEEE/AIAA 34th Digital Avionics Systems Conference (DASC), 2015, p. 8C6-1-8C6-12.

[113] G. Tsiligiannis, L. Dilillo, A. Bosio, P. Girard, S. Pravossoudovitch, A. Todri-Sanial, A. Virazel, H. Puchner, C. Frost, F. Wrobel and F. Saigné, "Multiple-Cell-Upsets Classification in commercial SRAMs," IEEE Transaction on Nuclear Science, DOI 10.1109/TNS.2014.2313742, 2014.

2

RHBD Techniques for Memories

Cristiano Calligaro

RedCat Devices, Italy

"Tell me about the waters of your birthworld, Paul Muad'Dib."

Frank Herbert, Dune

Radiation Hardening by Design (RHBD) may be considered as a set of techniques used to mitigate radiation effects on silicon devices. Since semiconductor components may be different according to the technology used (CMOS, Bipolar, BiCMOS) and the application they target (analog, digital, mixed signal, memories), RHBD techniques may be adapted and applied to fit requirements coming from the space domain. Semiconductor memories are quite peculiar from the design point of view (they are mostly digital in the periphery but strongly analog in the sensing schemes), and this peculiarity requires a specific set of rules to mitigate both soft and hard errors. In this chapter, we will discuss the main techniques used for both volatile and non-volatile semiconductor memories.

2.1 Effect of HEPs on Semiconductor Devices

The generation of hole–electron pairs (HEPs) coming from the direct ionization of a charged particle (heavy ions) or the secondary nuclear effects from an impinging proton is the main threat in silicon devices.

The passage of the charged particle creates a region where holes and electrons are separated thanks to the energy release; the higher is the energy, the higher is the number of HEPs. These pairs of course try to achieve a recombination, and this is true if boundary conditions permit this natural event. But if an electric field is applied, then electrons may move in one direction and holes in the opposite direction (see Figure 2.1); the higher is the electric field, the higher is the number of HEPs that will not be recombined.

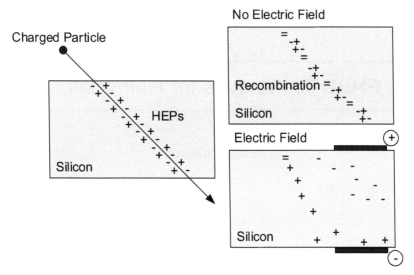

Figure 2.1   Effects of charged particle strike in silicon.

HEPs generation has different effects according to the place where they are generated. In those parts where doped silicon creates drain and source regions, the charged particle track creates the so-called "Funnel" region, a sort of cloud of electrons extending from the highly doped areas down to the bulk of the transistor (see Figure 2.2). The same funnelling effect can be observed from Deep N-well regions (if available in the CMOS process) into P-doped regions.

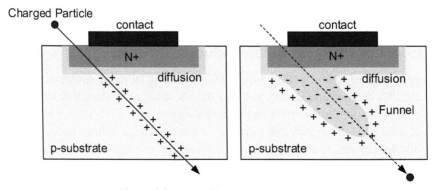

Figure 2.2   Funnelling in silicon devices.

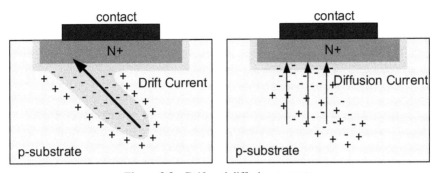

Figure 2.3 Drift and diffusion currents.

Funnel regions are the media through which currents are created. The energy release of the charged particle creates a drift current, while the natural recombination of holes and electrons create a diffusion current (see Figure 2.3). The two components are different in times (drift currents are generated in few tents of picoseconds, while diffusion currents are over after a couple of nanoseconds), but the overall result is that energy release has been transformed in a current spike responsible for a charge injection.

HEPs in silicon dioxide have a different effect. If in doped silicon holes and electrons are recombined or migrated to positive/negative terminals in dioxides, HEPs have to consider mobility issues during migration if an electric field is applied. Electrons, thanks to their size, can move with a higher mobility than holes and this takes to a natural recombination where electrons are the players, but the higher mobility can also distract them if an electric field is applied and this agility takes them far from their half of the pair. In this case, holes still move in dioxide to find a recombination, but in their path, they can find defects of the lattice able to stop them and trap. These are called charge traps, for which the manufacturing process of silicon dioxide (LOCOS or STI) is responsible. Having a clean oxide is difficult and requires a very fine manufacture, and even with high-scale technologies (e.g. 65, 45 or 28 nm), some defects are always possible.

In general, the thicker is the oxide, the higher is the probability to find defects and, of course, the higher is the probability a hole is trapped. The thinner is the oxide, the lower is the probability to have defects. This is the main reason why deep submicron ($<$100 nm) technologies are less prone to charge traps (but on the other hand more prone to transient, as we will see in the following paragraphs).

Whenever they are generated, silicon dioxide and doped silicon HEPs are responsible for errors in semiconductor components, which can be divided into two families:

a) Total Ionizing Dose (TID) and
b) Single Event Effects (SEE).

The first are those errors coming from the exposure of semiconductor components to radiation sources, and the second includes errors coming from specific particles impinging silicon devices. TID is a cumulative effect (the more prolonged is the exposure, the higher is the probability to fail), while SEE is stochastic.

Semiconductor components may have two different kinds of errors: a) Hard Errors and b) Soft Errors. Hard Errors cannot be recovered (like latch-up), whereas Soft Errors are those giving a temporary error such as a bit flip or simply a transient propagating in the chip and that can be recovered via software or passing through hardware correcting systems (EDAC, Error Detection and Correction). It is evident that Radiation Hardening by Design (RHBD) techniques must, above all, guarantee a mitigation of Hard Errors and then a reduction of Soft Errors.

2.2 Cumulative Effect: TID

Standard CMOS transistors (both n-channel and p-channel) use silicon dioxide widely in the overall front end of the chip. Silicon dioxide is used to isolate polysilicon gate from transistors channel and to isolate contiguous differently biased transistor. In all cases, the key role is isolation, and the better is CMOS manufacturing process, the better is the behaviour of the transistor itself and its independence from any biasing coming from neighbour transistors.

Silicon dioxide manufacturing is one of the key processes in a silicon foundry CMOS flow and relevant efforts making pure oxides have been made in the past to obtain a low number of defects both in thin (channel) and thick (isolation) using different techniques (e.g. dry oxidation).

In old manufacturing process, isolation between transistors was guaranteed by Local Oxidation of Silicon (LOCOS), where the passage from thin gate oxide to thick oxide was gradual and shaping the so-called "bird's beak".

With modern technologies, a different technique, called Shallow Trench Isolation (STI), overrode LOCOS, making more scaled and efficient isolation between contiguous transistors.

Charge trap phenomena are almost the same in both gate oxides and isolation oxide, and the effect is the same for both LOCOS and STI.

In Figure 2.4 is sketched a typical n-channel transistor in a CMOS process using STI. Drain and source regions (collectively called Active Area or Active Region) are surrounded by silicon dioxide to guarantee isolation from other transistors biased at different voltages. Even if CMOS is a planar technology, nevertheless, STI thickness must guarantee the isolation blocking any biasing path passing through the bulk (p-substrate) of the chip. Thick oxide means a higher probability to have defects and, consequently, to give rise to charge traps in case of holes generated by charged particles.

The same may occur in thin gate oxide (below polysilicon) responsible for the modulation of gate to create the channel path for the passage of charges from source to drain, but considering this time the oxidation is thin, a lower number of defects are expected, thus decreasing the probability to have charge traps. Deep submicron CMOS processes (below 100 nm) are

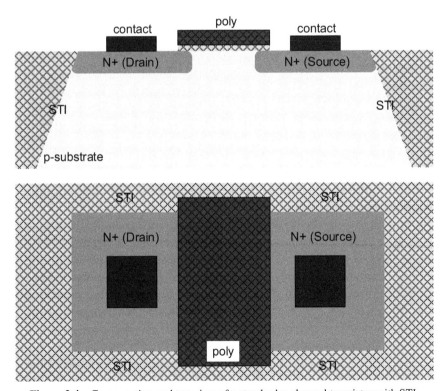

Figure 2.4 Cross section and top view of a standard n-channel transistor with STI.

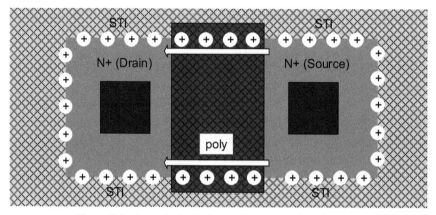

Figure 2.5   Intra-device leakage effect on transistor borders.

considered by the Space Community safe enough to consider any TID effect negligible at transistor channel level. Of course, the same cannot be said for STI; considering the thickness, they still have to guarantee isolation between neighbour transistors.

Since charge traps are responsible for leakage paths, we divide TID effects into two: (a) Intra-device leakage and (b) Inter-device leakage.

Intra-device leakage affects the single transistor only shaping its width and length and modifying its general performance; Inter-device leakage affects different transistors for a biasing of the first affecting the performance of the second (and vice versa). Inter- and Intra-device leakage may also combine their effects leading to a further degradation that, it is worth highlighting, is not recoverable (Hard Error).

Figure 2.5 shows the effect of charge trapping in STI isolating a single n-channel transistor in a typical CMOS technology.

The progressive effects of HEPs generation with time (month after month and year after year) lead to a collection of positive traps (holes trapped by defects in STI) in proximity of the interface between silicon dioxides and active areas. This effect is practically negligible in all drain and source regions except those corresponding to channel transistor borders. In the middle of the channel, there are not relevant effects, but where thin channel oxide becomes thick isolation oxide (Bird's Beak or STI), a conductive path can bring charges from source to drain overriding the role of transistor gate.

From the electrical point of view, the overall effect is a deviation of the threshold voltage of the transistor. In the middle of the channel, the threshold is only partially or negligibly affected by charge traps, while in the borders,

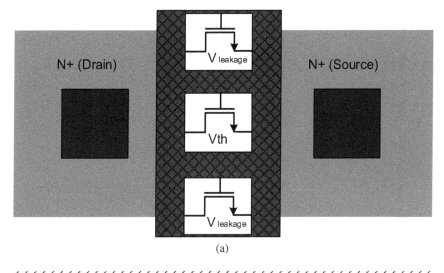

(a)

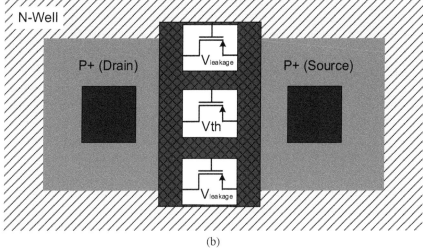

(b)

Figure 2.6 (a) Threshold deviation in an n-channel transistor due to Intra-device leakage. (b) Threshold deviation in a p-channel transistor due to charge trapping.

the effective threshold is decreased to a lower value; the overall threshold will be the medium value of the three contributes. Once the leakage paths on border will be strong enough to show $V_{leakage} < 0$, the transistor will not be able to switch on and off its status: it has become a leaky transistor and its operations are compromised [1] (see Figure 2.6(a)).

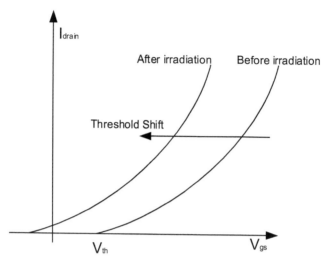

Figure 2.7   Threshold shift of an n-channel transistor in pinch-off region ($V_{ds} > V_{gs} - V_{th}$).

Concerning p-channel transistor, the situation is the same in terms of charge trapping but opposite in terms of threshold shift.

Along transistor channel borders, positive traps populate the region where thin oxide becomes thick oxide, but since this time, along the channel, holes move from drain to source instead of electrons, what a conductive path in the n-channel becomes a resistive path in the p-channel. In this case, the threshold of parasitic transistors on the borders will increase and current flow will become lower; when threshold voltage overrides the power supply, transistor will be always in cut-off (see Figure 2.6(b)).

Figure 2.7 shows the threshold shift of an n-channel transistor working in pinch-off (or saturation) region before and after irradiation.

Inter-device leakage involves neighbour transistors, and the effect is the biasing between different terminals. Figure 2.8 shows an example of a pair of n-channel transistors separated (and isolated) by STI. These HEPs generated from the cumulative effect of charged particles hitting the device create charge traps that remain stable with their positive charge at the interface between silicon dioxide and p-substrate of the chip. The thickness of trench in modern CMOS technologies is enough to avoid any possible biasing in standard conditions thanks to the depth of the trench, but the growing number of traps at the interface creates a leakage path getting in communication N+ areas of different transistors. In case of a terminal having a specific voltage applied, the second may be affected. This phenomenon is particularly

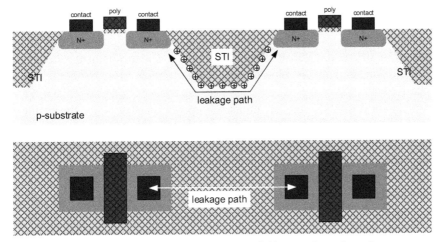

Figure 2.8 Inter-device leakage between neighbour n-channel transistors.

dangerous in analog electronics, where few millivolts can affect the working conditions of, for example, an operational amplifier or a comparator.

A similar situation can occur when an n-channel transistor is close to a p-channel transistor (see Figure 2.9). In this case, the leakage path can make a connection between the N+ diffusion of one of the terminal of the n-channel transistor and the N-Well containing the p-channel transistor. In this case, if, for example, the N-Well is biased at power supply, then the leakage path may bias the n-channel transistor terminal moving to a different voltage value.

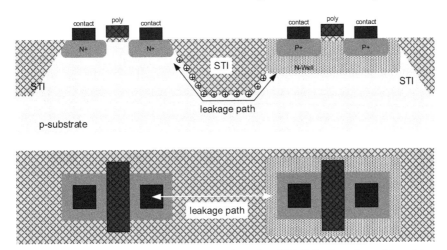

Figure 2.9 Inter-device leakage between n-channel and N-Well of a p-channel transistor.

2.3 Single Event Latch-up: SEL

In space components, SEL is considered the major threat and many efforts are made to make silicon devices resilient to this effect. The mechanism relies on vertical and lateral parasitic bipolar transistors coming from N-doped and P-doped regions.

In a standard CMOS process, parasitic bipolar transistors are the result of differently doped regions containing n-channel and p-channel transistors. P-channel transistors are those having drain and source heavily doped (P+), but since the substrate of the silicon wafer is lightly doped with the same dopant (P), it is necessary that they are included in a well-doped N: this region is called N-Well. On the other hand, n-channel transistors, having drain and source heavily doped N (N+), can be built directly on the p-doped substrate of the silicon wafer.

This sequence of N-doped and P-doped regions can build bipolar transistors. On the p-channel side, starting from P+ diffusion of a drain/source and passing through the N-Well, the P-substrate is reached, thus giving a P–N–P junction, also called vertical PNP bipolar transistor. On the n-channel side, starting from N+ of a drain/source and passing through the substrate, it is possible to reach the N-Well: in this case, there is an N–P–N junction, also called lateral NPN bipolar transistor. Moreover, this pair has two common nodes: the base of lateral npn connected to the collector of the vertical pnp (the p-substrate) and the collector of lateral NPN connected to the base of the vertical pnp (the N-Well). This connection shapes the typical positive feedback loop, and the structure is also called SCR (Silicon Controlled Rectifier) or Thyristor (see Figure 2.10).

SCR is normally in off condition, and current flowing is negligible. This state is also called "blocking state", but it is possible to exit from this condition if a voltage is applied to the intermediate P terminal (node G in Figure 2.11): a sort of control gate. In this case, a current can flow

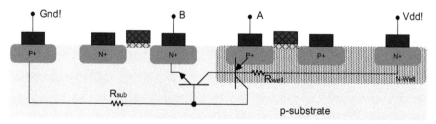

Figure 2.10   Silicon Controlled Rectifier (SCR) in a CMOS standard process.

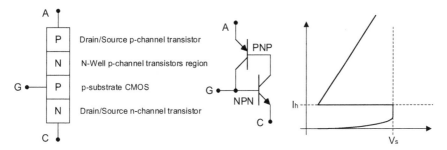

Figure 2.11 Silicon Controlled Rectifier (SCR or Thyristor).

from P-anode (node A) to N-cathode (node C) [2]. When triggered into the conductive state, the SCR enter the "latched state", which results from a current injection in the base of the NPN transistor (in Figure 2.10, this is the p-substrate). As a result, it creates a current flow in the base–emitter junction of the PNP transistor, which turns on and increases further the current injected in the NPN transistor: the positive feedback loop is triggered and both transistors are saturated; the current flowing in both transistors guarantee that SCR does not exit from the latched state.

To take back SCR in blocking state, the voltage across terminals A and C (see Figure 2.11) must be reduced to decrease the current flowing in vertical PNP and lateral NPN or, alternatively, the current flowing into the base of lateral NPN must be reduced as well. Once the current is below the holding value, the SCR is off.

In CMOS standard technologies, SCR is usually under control with a good power distribution in the chip and the only way to fall into the so-called "Latch-up" is through power supplies or ground. A spike on these terminals can take the parasitic SCR in the latched state, thus triggering the positive feedback loop and taking vertical and lateral bipolar transistors to sink current. The overall effect is the melting of metal junctions leading to a permanent malfunction.

In space applications, HEPs generation must be considered, and if charged particles releasing a low energy (low LET) are not responsible for any destructive error, those particles releasing high levels of energy (high LET) can produce a deep funnel region involving both p-substrate and N-Well. The generation of an energetic funnel (see Figure 2.12) produces a drift current followed by a diffusion current, and if the ground strategy of p-substrate is not strong enough, a resistive path (distributed parasitic resistance) can be on the road of the funnel. In this case, the drift current, usually involving the

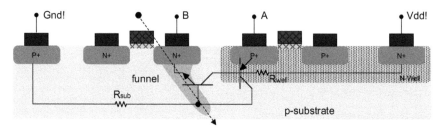

Figure 2.12 Single Event Latch-up (SEL).

diffusion of the transistor terminal where it comes from, can provide part of its current in the base direction of the lateral NPN transistor, thus increasing the current flowing in the vertical PNP transistor. In this case, a single current spike overriding the hold current of the parasitic SCR takes the region in the latched state and current from power supply takes advantage of the positive feedback loop of the SCR causing the latch-up.

2.4 Single Event Upset: SEU

HEPs generation at lower energy (low LET) generates funnel regions not affecting substrate and N-Well (the funnel remains in the region of N+ or P+ diffusion) but still can create drift currents. Such current spikes can create transients (Single Event Transient, SET) consisting in the charge of parasitic capacitances around the N+/P+ diffusion area.

Figure 2.13 shows the small-signal circuit for an n-channel transistor in a standard CMOS process [3]. Considering the four-terminal mode (Gate, Drain, Source and Bulk), different capacitance and resistive contributions come from different junctions. Starting from the bulk (terminal B), there are R_{BS} and R_{BD}, which are the distributed resistance connecting the substrate to source and drain regions; C_{TS} and C_{TD} are depletion-region capacitances at the reverse-biased p–n junction between substrate and source and substrate and drain, respectively. R_S and R_D are the resistances coming from the contact terminal (drain and source) and the channel of the transistor. On the gate side, there are four parallel capacitances: C_{GD} and C_{GS} are inherent to the transistor as a device, while C_{RD} and C_{RS} are parasitic elements coming from the manufacturing flow and are becoming lower and lower thanks to scaling of the technology node and more precise masking tools. C_{GD} and C_{GS} are the capacitance between gate and drain/source terminals and are strongly dependent on the bias condition of the transistor and its width and length.

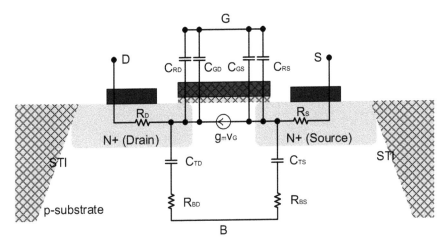

Figure 2.13 Small-signal equivalent circuit for a CMOS n-channel transistor.

C_{RD} and C_{RS} represent the capacitance coming from the overlap between the gate and drain/source diffusions.

According to the small signal equivalent circuit, it is now possible to evaluate the effect of a charge particle striking even at low energies (low LET). The funnel region creates a drift, and immediately after diffusion, a current basically flowing in the N+ region (see Figure 2.14).

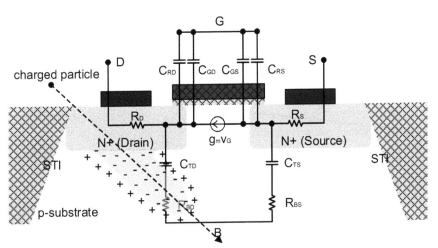

Figure 2.14 Charge particle strike in a CMOS n-channel transistor.

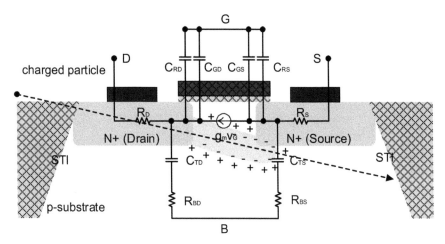

Figure 2.15   Charge particle strike generating a funnel region including drain and source areas.

The current generated sees several capacitances on its path: C_{TD}, C_{RD} and C_{GD} (according to the example shown in Figure 2.14). Being a low energy strike, the contribution of C_{TD} is negligible but C_{RD} and, in particular, C_{GD} must be considered. The charge (or discharge of these capacitance) is responsible for the transient on node D; if this node is well biased, the effect is negligible, but in case of a specific connection (e.g. a bistable circuit or a diode connection in a current mirror), the result can be an upset or an amplified current spike triggering a cascade effect in a digital flip-flop or an analog component.

The incident angle of a charged particle is a stochastic event and the hitting may involve more diffusion area, as shown in Figure 2.15. In this case, the drift and diffusion current may come from two parts and involve more capacitances, giving more complex charge and discharge phenomena capable of triggering undesired transients.

Figure 2.16 shows the hitting of a particle strike on one of the two bistable nodes of 6T SRAM cell. These nodes are opposite to each other and can be only at ground or power supply. The great advantage of this circuital topology is that in both states, the structure has no current flowing along the p-channel and n-channel transistors. The only currents flowing are those able to compensate the leakage currents from transistor junctions, which are negligible in the overall economy of the device. The stability of the bistable pair is guaranteed by capacitors (see Figure 2.13), and to change the state,

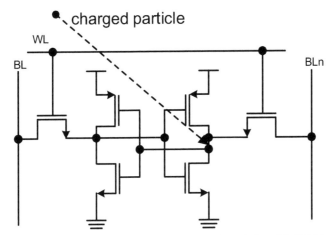

Figure 2.16 Charged particle strike on a bistable node in a 6T SRAM cell.

a voltage can be applied to one (or both) node to make the bistable flipping in the opposite direction. This voltage charges capacitors, and this charge is also called Critical Charge. Once overrode the critical charge, the bistable circuit, thanks to its positive feedback loop, changes state.

A charge particle releasing energy generates drift and diffusion currents; if these currents can charge capacitances keeping the bistable in the previously programmed state overriding the critical charge, we have an upset (a logic "0" becomes a "1" or vice versa); if the currents are not enough, then we simply have a transient.

2.5 From SET to SEU/SEFI/MBU: When a Disturbance Becomes an Error

As mentioned in the previous paragraph, the energy released by an energetic particle becomes a charge injected in transistor nodes having intrinsic capacitance related to the device itself or misalignment of the masks in the manufacturing flow. This event is also called fault injection, and the basic assumption is that every node in the net has a specific critical charge that we can also see as the capability to absorb the energy released by the particle. For very low energetic particles, the effect is negligible, but when the LET increases, such nodes start showing glitches.

In standard logic ports such as inverters, NANDs or NORs, a particle strike can produce a temporary (and very short in time) voltage spike, which can be considered as a disturbance, and if the architecture of the logic device is robust enough, the disturbance may be filtered by using the same methods commonly adopted in digital electronic devices (filters, good disoverlap strategies, etc.). When energy becomes higher, the spike can become relevant in terms of dynamics and the output signal (see Figure 2.17) can change state for a while before recovering to the original value.

According to the amplitude of the transient, the SET can remain in the port hit by the particle or it can propagate to the next cell (see Figure 2.18). In this case, it is possible to have a cascade of glitches running through logic into the critical part of the chip.

Figure 2.19 shows how an SET originated by a charged particle can run into a logic net (a set of simple logic ports) and once arrived, the reset terminal of a latch can change a logic value, which is expected to remain high according to the logic function of the digital block. The SET has become an SEU, and a simple disturbance has been transformed into a soft error.

The example of Figure 2.19 can be moved to more complex architectures, and the soft error can make a second cascade involving more storage elements shaped as a state machine. In digital design, it is quite common to have state machines programmed to have a specific number of states where functions are performed; sometimes not all possible states are considered (to reduce the

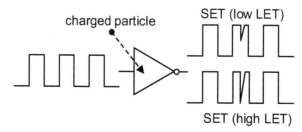

Figure 2.17   Charged particle strike in standard CMOS inverter.

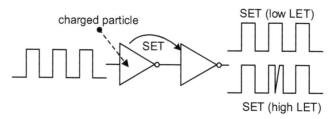

Figure 2.18   SET propagation in a logic chain.

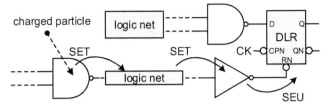

Figure 2.19 SET propagation into a Delay Latch with Reset (DLR) becoming an SEU.

number of digital cells), which are known as Finite State Machines (FSM). An SEU can take the FSM in a forbidden state where only a reboot or a restart of the digital engine can take out from this stuck condition. This kind of error is also called Single Event Functional Interrupt (SEFI).

Figure 2.20 shows an SEFI occurring to a counter reading a memory (SRAM or PROM) due to an SEU on the Most Significant Bit (MSB); in this case, if the particle strike comes when MSB is "0" taking it to "1", a relevant number of addresses are skipped and the loaded code is wrong. On the other hand, if the hitting takes from "1" to "0", the counter comes back to a previous address and reads again the same set of addresses giving in any case a failure in the code loading procedure. In this case, even if the digital

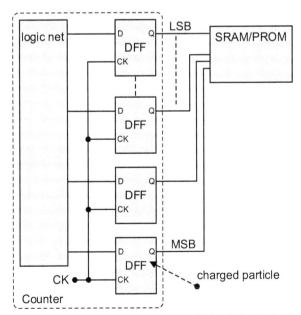

Figure 2.20 SEU in counter scanning a memory (SRAM/PROM) for code loading.

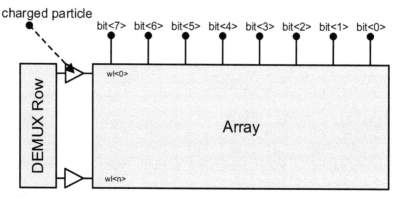

Figure 2.21   MBU from a charged particle hitting a row decoder final stage.

engine does not enter a forbidden state and stuck condition is not reached, a functional interrupt (SEFI) comes from the wrong code loading.

In semiconductor memories, the array containing memory cells (6T SRAM or non-volatile) are usually organized to minimize area consumption; this means that along a word-line, many cells (bits) share the same wire. Figure 2.21 shows a sketch of a typical memory organization. When the memory cut is small enough, all the bits building the byte share the same word-line and final stages (buffers) are usually sized to drive efficiently both the closest and the farthest bit minimizing any delay. In this case, the wrong decoding in the DEMUX stage or the direct hitting of a charged particle on the terminal stage driving a word-line can make a mistake along all the bits: this is called a Multiple Bit Upset (MBU).

In large memories, the array can be split and the common DEMUX engine drives two word-lines (left and right) with separated drivers (see Figure 2.22).

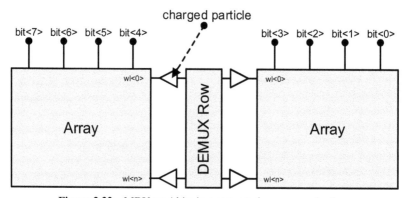

Figure 2.22   MBU on 4 bits in a separated array organization.

In this case, if the charged particle hits a driver, the result is that 4 bits are wrong, but still if the particle hitting occurs on the DEMUX engine, a wrong word-line can be decoded, thus leading to an MBU on all the bits of the byte.

2.6 Radiation Hardening By Design (RHBD)

Radiation Hardening by Design is a set of techniques focussed on the mitigation of the effect of radiation on semiconductor components. These effects cannot be completely avoided, and also by using heavy shields, some effects still pass. Just like occupational health and safety for men at work, also for semiconductor components in hostile environment, the risk of a failure can be only reduced at its minimum but not completely avoided. So, like men at work in a building site, even if protected by the right equipment and following proven procedures, semiconductor devices can fall into an accident.

RHBD approach is very similar to occupational health and safety procedures. Here, the main task is to reduce the probability of an accident at its minimum, keeping in mind that it can always occur even in the best conditions; the second effort is to reduce at its minimum the effects of an accident in case it happens. RHBD techniques first of all face the destructive events (SEL, TID) and reduce the probability that they can occur with architectural strategy and by using a dedicated physical design; once reduced the risks of such events, the second task is to mitigate the effects at its minimum.

There is a third task very well known in the space community that is represented by EDAC (Error Detection and Correction). It is a sort of a cure of radiation effects, but it is not (and must be not) considered as a panacea. As always happens, EDAC gives the best when a solid architecture behind and good layout design give few errors to be corrected; pretending EDAC resolving all issues coming from components not conceived to be rad-hard is risky.

RHBD has been considered in the past as a handcrafted activity inside the semiconductor market. Nowadays, this is only partially true.

The progressive scaling down of modern technology is taking transistors in the few tents nanometre area with very reduced power supplies and, of course, very low critical charges in bistable elements. Also, hard errors may occur more frequently, and relevant efforts are spent in power strategies and solid ground connections. In the nanometre era, semiconductor devices may be prone to particles with energies much lower compared to those space community have to consider (heavy ions, protons, gamma rays), and as a matter of fact, aircraft producers are starting to make strategies for electronic equipments able to resist some kinds of radiations. This cannot still be considered as a rad-hard or rad-tolerant approach but may be highly reliable: the threat behind is very similar.

Radiation hardening can be done not only by using a dedicated design but also by making leverage on the manufacturing process or by using dedicated shielded package. The first is called RHBP (Radiation Hardening by Process), and the latter is called RHBS (Radiation Hardening by Shielding).

RHBP represents a very appealing option and indeed is used for some very specific applications. Making a dedicate CMOS or BiCMOS with additional masks or specific manufacturing steps can be a way to solve the root cause of HEPs generation effects from the viewpoints of both TID and SEE, but it has several drawbacks. The first is of course the cost: making a dedicated process for a very low number of components (as usually requested by space market) can be a waste of resources that can be justified only by classified applications (military) or very specific and critical tasks.

But in addition to the cost, there is also another problem in the RHBP approach. Creating a new "space silicon process" can be feasible but "maintaining a space silicon process" is very difficult and can be the real killer task. It is well known that silicon manufacturing takes advantage of the number of wafers produced every month; new technology nodes, in the beginning, presents a high variability in the process and it is only with a fine tuning of the manufacturing steps that yield rises up to 99% as expected in a fine-tuned process. The higher is the number of wafers produced, the higher is the control the silicon foundry has on the process: this is called maturity of the process.

Now, a dedicated silicon process, like those done in the academy or research centres for exploring some new limits of the manufacture, cannot reach this kind of maturity because the number of wafers will always be low. Space components are sold not in millions of parts but in hundreds or thousands.

Space applications, nevertheless, need a mature process. It is well known that space community is quite conservative and usually not keen to adopt new technologies. This is also true for silicon process. The maturity of a process is something very welcome from rad-hard component designers or integrators. In some cases, producers prefer using existing components proven to have a very high reliability and shield them as much as they can. This is RHBS, and the effort is focussed on the package that must be light (to meet weight requirements) and able to reduce the LET of charged particles.

RHBS does not provide as a result a rad-hard component, but it is usually called a rad-tolerant part, and even if they do not reach the same resistance of rad-hard components, they play a significant role in the space market. Those missions in low orbit (LEO) or missions having a very limited time frame can take benefit of rad-tolerant components that usually have a lower cost and, on the other hand, a higher number of functions (or storage capacity if they are memories). As usual it depends on the mission and making a trade-off is the task that engineers have to take into account.

RHBD in the last 20 years has become one of the main assets for space components. The main pillar behind this approach is the use of standard silicon technologies; this means CMOS for standard logic, memories and digital ASICs and BiCMOS (SiGe) for advance analog components or RF devices.

As mentioned above, maturity is critical for space applications, and this is the reason why in RHBD selected process are usually not the most updated but part of the previous waves. Even if very aggressive nodes nowadays guarantee a strong minimization of area (and power consumption), nevertheless older technology nodes are preferred. Today, an "old" CMOS 180 nm or 130 nm can be beneficial for more than 20 years of continuous refinements and the high number of wafer manufactured every month lead to yields very close to 100%. This is definitely a positive aspect for applications asking, first of all, of a high reliability.

Behind the maturity of a process there is another positive aspect. These old nodes are available from many silicon foundries in all around the world, thus opening the possibility to move from one silicon foundry to another one.

Space components market represents a thin slice of the overall semiconductor market, but the main characteristic is to be worldwide; that means few components are sold every year and spread almost everywhere.

As a matter of fact, each country has its favourite silicon sources also in consideration of the fact that large programs are triggered by public entities, which are called to use internal sources; if a country has its own silicon foundry, then there is a high probability that it will be the provider in the manufacturing flow.

So, the portability becomes critical. A space component designed in a CMOS 180 nm and provided to a country must be moved to a different CMOS 180 nm in another country and if this can be considered an obstacle, on the other hand, it takes the overall approach under "safe" condition.

The passage of a space component from a silicon source to another reduces the risk if a silicon foundry dismisses a process or decides to stop the manufacturing for that market.

Also national rules (like ITAR, International Traffic Arms Regulation) can be considered a threat for space components and if a specific device today is manufactured with a standard CMOS process from a foundry that tomorrow is acquired by a larger corporate that decides that this process becomes classified, the risk is to be not able to have a manufacture anymore.

RHBD represents an umbrella for space components. If well used, it can enable business opportunities, but it can also protect by possible changes from silicon sources also considering that space community use these components, once qualified, for many years: what it can be profitable for a foundry today (because in parallel of space components there are other consumer applications) can become an obstacle tomorrow.

Under these premises, RHBD and in consideration of the main pillars described above (Maturity and Portability), it is now possible to enumerate some possible steps for making space components rad-hard starting from a standard silicon process.

Starting from the silicon foundry Process Design Kit (PDK), the first thing to be done is to discard everything that has been designed and that is part of the PDK: from standard cells [4] to I/Os. This can be considered a radical approach, but the main assumption is that everything has been designed by process engineers not for a rad-hard application but for different purposes, and only one failure can (e.g. a latch-up in an output pad) frustrate an entire chip.

A PDK containing only Design Rule Check (DRC) and Layout versus Schematic (LVS) engines is also called naked-PDK (nPDK), and this is the starting point for making a real rad-hard component.

A good and robust architecture is the starting point for almost every rad-hard component; the right macro-blocks positioning, redundancy and a preliminary floor-plan to guarantee an efficient power distribution and ground lines is the main brick for making easier any successive step both from the topological point of view (the circuits) and from the physical layout point of view. This is called RHBD at Architectural Level (RHBD-AL).

In the top-down approach, as sketched in Figure 2.23, the next step is represented by the circuit design where topological solutions focussed on minimizing the impact of charged particles are considered to avoid hard and soft errors. This is called RHBD at Circuit Level (RHBD-CL).

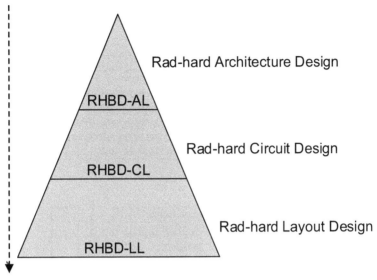

Figure 2.23 RHBD top-down approach.

The final step is represented by the physical layout design. A strong architectural approach combined with robust circuit solution paves the road to rad-hard devices, but the layout plays a crucial role since the main hard errors coming from TID and SEL are solved with a good and reliable design. This is called RHBD at Layout Level (RHBD-LL).

2.7 RHBD at Architectural Level

A generic semiconductor memory (volatile or non-volatile) is built by specific block shaped to be digital or analog. In general, the main block is the array containing the memory cells (bistable elements for SRAMs and Flash for non-volatile), while all the circuitry driving the array is called periphery.

Figure 2.24 shows a generic non-volatile memory diagram; DEMUX for rows and columns are strictly digital parts; sensing usually has a design closer to analog devices, while High-Voltage generators (charge pumps) provide the right conditions to program non-volatile memory cells. In a non-volatile memory, programming and erasing are usually driven by an algorithm working in parallel with high-voltage generators and sensing circuits to implement the so-called Progam&Verify or Erase&Verify mechanisms. These algorithms make large use of state machines.

From an RHBD perspective, all these blocks must evaluated in terms of hard/soft error, making a sort of risks matrix.

Macro Block	Error	Description
Input/Output Driver	SET	Risk of SET cascade
HV Generator	SEL	Charge pumps are prone to SEL
P/R Algorithm	SEFI	FSM can provide wrong algorithms
DEMUX	SEU/MBU	Risk to decode the wrong row/column
Array Driver	SEU/MBU	Final stages can select wrong row/column
Array	MBU/MCU	Memory cells can be upset

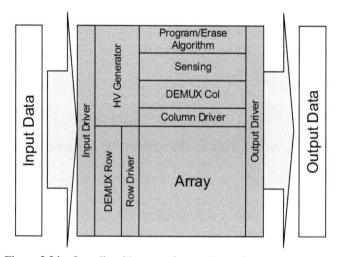

Figure 2.24   Overall architecture of a generic semiconductor memory.

From the above table, it is possible to split the architectural effort into two ways: (a) the matrix, where multiple errors must be reduced, and (b) the periphery, where the main threat comes from wrong row/column decoding that can lead to MBU.

SEL, as a hard error, affects all blocks since in all parts n-channel and p-channel transistors can be close enough to trigger the latched state of parasitic SCR. Nevertheless, HV generator can be an additional threat, considering that charge pumps are particularly weak, in particular, when they include n-channel transistors.

Figure 2.25 shows an RHBD-AL approach to a generic non-volatile memory. Since the erroneous decoding of word-line or bit-line can lead to multiple errors, the first attempt is to reduce MBUs to SEUs. A row driver

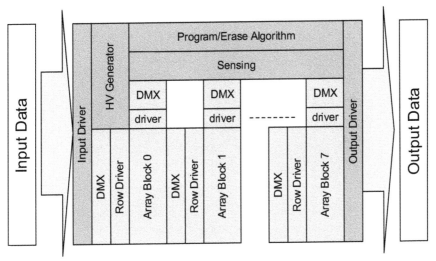

Figure 2.25 RHBD-AL on a generic non-volatile memory.

failing can be replaced with a row driver for each bit; in this case, a charged particle hitting a driver will have no effect on the other and the failing bit will be only one. The same approach is valid for column drivers.

The independence of separated matrix arrays on their drivers is a first step, and the same approach can be adopted on both DEMUXs for rows and columns, but in this case, the redundancy, beneficial to reduce further MBUs, can become too much expensive in terms of area and power consumption. A possibility can be a redesign of a DEMUX having some precautions at circuit level (so demanding to RHBD-CL the problem) or make a mixed approach; if, for example, the number of bit-lines is lower than word-lines, then the column DEMUX can be split (one for each bit) keeping the row DEMUX in common for all the rows.

Another RHBD-AL technique can be a partial redundancy using voters. In this case, instead of replicating eight times a DEMUX, it is possible to use three DEMUX only and use a voting scheme: if two decoders among three define a row (because the third has been affected by a charged particle strike), then the result is strictly based on the majority of the votes (see Figure 2.27).

In the RHBD-AL area falls also the ECC approach. Choosing the right approach is critical because the waste of area is relevant. ECC algorithms, in fact, are based on an increased storage capacity where to store the data and the keyword able to tell if the reading operation has been carried out

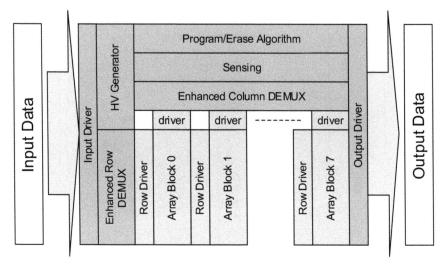

Figure 2.26   RHBD-AL with array and terminal drivers split.

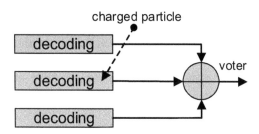

Figure 2.27   Decoding scheme based on voter.

without errors. In addition to error detection, more sophisticated engines can also correct the error but of course requires additional storage capacity and a good algorithm that can implement this operation during writing and reading.

ECC algorithms may be SED (Single Error Detection), SEC (Single Error Correction) SECDED (Single Error Correction and Double Error Detection) and so on. The higher is the capacity to detect and correct, the higher is the complexity of the algorithm and storage capacity to implement it.

In an RHBD-AL approach, architecture and ECC must be carefully balanced. In a memory like the one described in Figures 2.25 and 2.26, where a first RHBD has been done to reduce MBUs to SEUs, maybe only an ECC implementing an SEC is enough because the first choice makes confident the component to reduce the probability to have multiple errors that can come not directly from decoding schemes but indirectly from the other blocks (Input

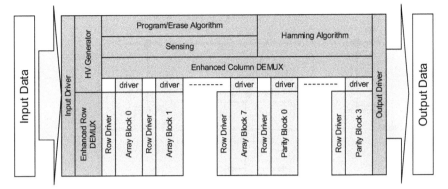

Figure 2.28   RHBD-AL combining ECC and redundancy.

Drivers). As mentioned in Chapter 1, the possibility that an MBU occurs is not null but it has been mitigated at its minimum.

An ECC engine implementing an SEC algorithm can be a 4-bit Hamming Code. In this case, in addition to the 8-bits data, 4 bits for parity must be considered together with encoding and decoding schemes.

Whenever it is redundancy or ECC (or both as shown in Figure 2.28), the designer must be aware that the main impact is an increase of area and power consumption. Both in stand-alone and embedded memories, the penalty to pay is relevant and the selected architecture must fit the requirements coming from the mission. Making an architecture with high redundancy and complex ECC for a satellite working in low orbits can be a waste of resources. Moreover, ECC are usually the field where core processors already play a role in the overall economy of digital ASICs; providing a memory having ECC on board can collide with ECC strategies of the final application. In this case, it could be preferable to provide simply storage capacity and leave all ECC issues to the digital ASIC implementing EDAC engines at a higher abstraction level.

2.8 RHBD at Circuit Level

RHBD at Circuit Level is the typical domain of microelectronic designers dealing with single transistors or standard logic cells considering topologies and structures to enhance resistance against radiations. In this domain, the major part of the efforts is devoted to mitigate the effects of transients, avoid SET propagations and reduce the probability to generate SEUs.

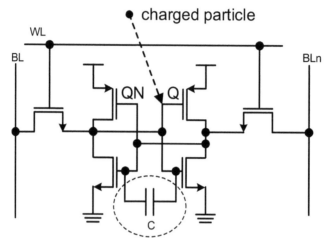

Figure 2.29   RHBD-CL on a 6T SRAM cell.

Figure 2.29 shows a typical RHBD-CL technique for increasing the stability of a 6T SRAM cell when a charged particle strikes one of the two bistable nodes.

As mentioned in the previous paragraphs, the node involved in the hit (Q) has a critical charge mainly coming from gate capacitance and junction capacitances. The positive feedback loop keeps Q and QN in their opposite states, but if the injected charge is higher than the critical charge, then the feedback loop takes the nodes in opposite direction, thus generating a bit flip.

Since it is a matter of critical charge, one possibility is to increase it by adding a capacitor on both nodes. The capacitor placed on the pairs is called Miller Capacitor and can be built using additional polysilicon structures (Poly Capacitor) or different metal levels (Fringe Capacitor). The result is that a higher energy will be necessary to trigger the SEU.

The same approach can be used also for Latch and Flip-Flop. Figure 2.30 shows that a D-Latch, having a critical charge represented by the input capacitance of an inverter and the output capacitance of a CMOS switch, can be increased placing a capacitor in the positive feedback loop. This strategy is easier in latches and flip-flops more than in 6T SRAM cells because it is a matter of adding a capacitor possibly shaped as a standard cell. In this case, areas are quite different (6T cells must be at the minimum to build efficient memory arrays) and a D-Latch like the one in Figure 2.30 can be large but the benefit is that with a good rad-hard standard tool-set and

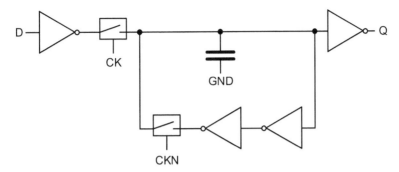

Figure 2.30 RHBD-CL in a rad-hard D-Latch.

capacitors placed to cut the positive feedback loop, it is possible to obtain structures resilient to SEU up to 60 MeV*cm^2/mg (the energy released by an impinging xenon ion).

Concerning SETs, they are the typical disturbances occurring to logic ports that, since they do not store any data, can only produce a voltage spike. RHBD-CL can mitigate the amplitude of such spikes and avoid their propagation along the chain of standard logic cells.

Figure 2.31 showns a typical XOR port using a 6T architecture, which is typical of almost standard cells. In this case, if a charged particle hits one of the transistors not directly connected to power supply or ground, the reduced dynamics of nodes can amplify the SET, increasing the probability of a propagation in the cell and out of the cell.

The XOR shown in Figure 2.31 is not rad-hard and placing capacitors on critical nodes to enhance critical charges can represent an overload for a cell not having storage capabilities and with the risk to reduce delays in the propagation of the signal.

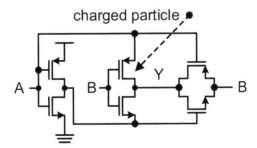

Figure 2.31 XOR logic port using 6T architecture.

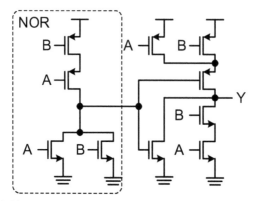

Figure 2.32 RHBD-CL in an XOR logic port using 10T architecture.

A possible solution is shown in Figure 2.32, where in a 10T architecture, the number of transistors not directly connected to the ground and power supply is reduced to its minimum. In this configuration, the probability to have an SET on such nodes is reduced and if it occurs, the propagation is prevented.

From the 10T XOR shown in Figure 2.32, it is possible to see that this port is the ensemble of an NOR with another elementary cell specific for the XOR transfer function. This approach is quite common in RHBD-CL, where the split of functions guarantees an increased resistance to SETs and their propagation.

Figure 2.33 shows a typical OR port with three inputs.

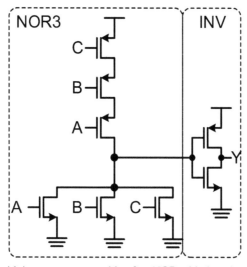

Figure 2.33 OR with inputs as an ensemble of an NOR with three inputs and one inverter.

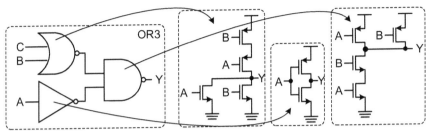

Figure 2.34 RHBD-CL in a three-input OR cell as an ensemble of NOR, NAND and inverter.

The basic block is represented by an NOR cell with three inputs with an inverter in cascade to change the output to the right signal requested by the OR function. This cell suffers of the same problem of 6T XOR port since there are two p-channel transistors not directly connected to power supply and where a charged particle strike can produce SET issues.

Keeping in mind the general rule of maximization of transistor terminal to ground (n-channel) or power supply (p-channel), a possible solution is the scheme shown in Figure 2.34.

In this case, the use of elementary cells having, all together, only two transistors (one in the NOR and the other in the NAND) not connected directly to power supply and the ground is beneficial with respect to the two p-channel transistors in cascade to each other and not connected to power supply.

The ensemble of Figure 2.34 takes 10 transistors (four for NAND and NOR and two from the inverter), while the original OR takes eight transistors: the penalty for a rad-hard architecture is two transistors, but the result is a cell resilient to SETs and mitigating propagation of disturbances.

2.9 RHBD at Layout Level

Layout domain is most critical in the process of making a rad-hard component, and even with a good architecture and a wise topological strategy for single circuitry, the role of a good physical design is crucial. In particular, hard errors are the ones to be considered at layout level and we include those coming from TID (degradation of transistor channel) and SEL (parasitic SCR triggered by charged particles).

Concerning SEL, the RHBD-LL shown in Figure 2.35 is one of the most efficient in avoiding the activation of the positive feedback loop represented

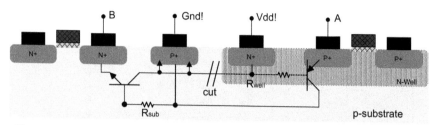

Figure 2.35   RHBD-LL SCR positive feedback loop cut by using enhanced guard ring.

by the parasitic SCR in CMOS technology in those regions where n-channel transistors and p-channel transistor are close to each other. In this case, just by placing substrate biasing and N-Well biasing in the middle of these regions is enough to create a cut in the feedback loop. The overall result is that the collector of lateral NPN parasitic transistor is connected to the substrate biasing (Gnd!) and the base of the vertical PNP parasitic transistor is shorted to the N-Well biasing (Vdd!). The reverse-biased P+ and N+ regions guarantee that there is not any path to close the loop and even in case of a funnel producing a drift current in the base of the NPN lateral transistor, this is discharged in the substrate-biasing terminal connected to the ground. The parasitic SCR cannot enter the latched stated, as better described by Figure 2.36.

From the design point of view, under these premises, layouts must be designed separating and isolating n-channel regions from p-channel regions (see Figure 2.37).

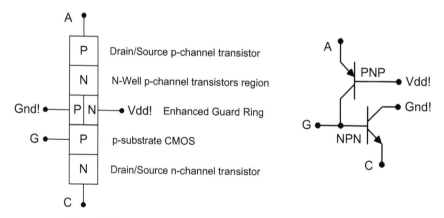

Figure 2.36   Role of the enhanced guard ring in the parasitic SCR.

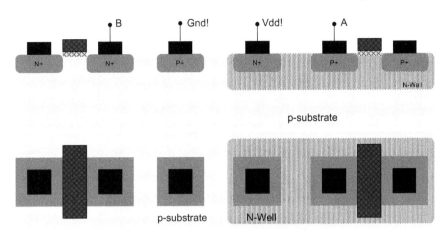

Figure 2.37   RHBD-LL technique to avoid SEL.

This technique has the great advantage to mitigate also TID from the inter-device leakage point of view in consideration of the fact that guard rings prevent any conductive path built by charge trapped in STI.

The second problem that can be solved only at the layout level is the degradation of the transistor channel along borders. Figure 2.38 shows a common structure used to avoid borders along the channel; this configuration is known as Edge-less Transistor (ELT) or Annular transistors, and it is shaped as a ring having one terminal inside and the second terminal outside. The result is still a transistor, and the poly line moving out transistor region onto STI still produces a leakage path, but since it is between two regions having the same potential, the result is null. This simple trick makes the transistor never leaky even at very high doses and it is the only way to maintain the stability of the threshold in every condition.

The intrinsic robustness of ELTs has several drawbacks: on top of all, the width that cannot be at the minimum provided by the lithography. Since the ring must be closed around the internal contact, the width will be always very far from its minimum and the same can be said for the length; the angle that in almost all technologies cannot be 90° but at least 45° (see Figure 2.38) produces a channel length not at the minimum of the node.

Another drawback of ELTs is that they are asymmetric devices: drains and source areas and perimeters are different, and this means that under some conditions, the device can work somehow differently from the canonical transistors. This effect is negligible if the use is digital (inverter, NAND,

NOR, etc.) but must be carefully evaluated in an analog design: making a current mirror, selecting a drain internally or externally in the ring may give a difference in the scaling (current multiplied by two or more).

The current flowing in the channel is not uniform like in standard transistors. Figure 2.38 shows a sketch of current contributions when moving from the drain to the source. In this example, also considering the positioning of the source contacts, the channel works in two directions (left and right) with

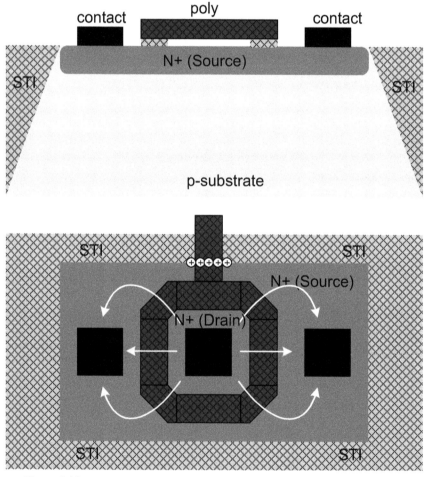

Figure 2.38   RHBD-LL Edge-less Transistor (ELT) to avoid border degradation.

contributions from the angles of the ring. In the top part (where the poly exits from transistor region), the contribution is very low and in the bottom part, if no contact is available in the source region, the same. Understanding the real width of an ELT is not easy, and most conventional CAD tools cannot determine the effective current flowing.

Figure 2.39 shows a design of an ELT with an additional contact in the source region. In this case, the channel works efficiently in three directions with minor contributions from the four angles. The fourth side (the one having the connection out of the transistor) still contributes partially to the overall current flow.

When transistor width is larger than the minimum, a good strategy is always to keep one side at the minimum admitted by the lithography and on that side to provide the poly connection of the transistor gate (see Figure 2.40).

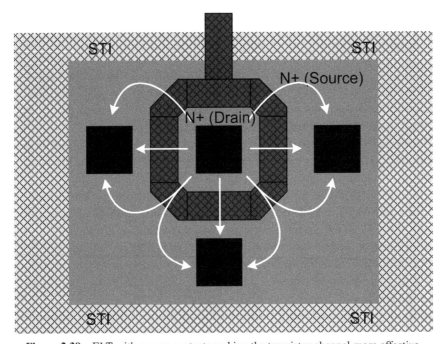

Figure 2.39 ELT with source contacts making the transistor channel more effective.

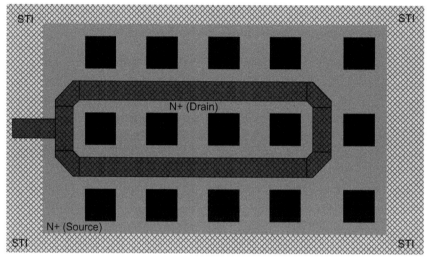

Figure 2.40   Contacts positioning in an ELT with larger width.

2.10 Conclusion

RHBD is a set of techniques to mitigate errors coming from charged particles affecting semiconductor devices. Such errors can be divided into hard (destructive) and soft (temporary) errors and can be the overall result of progressive exposure to radiation (TID) or stochastic (SEE).

Three different domains must be considered when designing a rad-hard component: (a) the architecture; (b) the circuit topology and (c) the physical layout.

Each of these domains mitigate specific effects. Architectural domain deal with SEFI and MBU, and only with robust main building blocks (e.g. FSM with all possible stated under control or bit distribution in a memory) it is possible to clean functional error bringing, in particular, in digital ASICs, in stuck condition. A wise architectural strategy makes easier, in the top-down approach shown in Figure 2.41, the following RHBD steps at circuit and layout levels.

RHBD-CL is in the middle between the architecture and the physical layout, and good circuit strategies can play a significant role not only in the circuit but also in the other domain. A robust circuit implementing an already robust function defined at architectural level makes the final device truly rad-hard not forgetting that a "simple" circuit can make the physical design process easier. RHBD-CL deals mainly with SET and SEU, but MBU mitigation and TID are also in the scope of this domain.

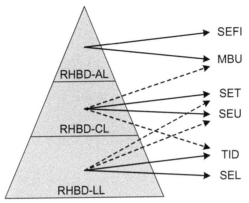

Figure 2.41   RHBD contributes to hard and soft errors in different domains.

RHBD-LL is the executive arm of architectures and circuits. Since everything must be, at the end, transformed in a physical layout, it is obvious that this domain plays a crucial role. The layout is 50% responsible for a successful rad-hard device, and this is because it is the one dealing mainly with hard errors. SEL, even if wise topologies are used, requires a robust layout and TID can be mitigated only with layout techniques (separation of n-channel and p-channel transistors, ELT).

Of course, layout cannot solve everything. Too much aggressive architectures (high-frequency clocks, FSMs not completely under control) and circuits (too much compact digital cells) cannot demand to layout only all radiation hardening issues; many compromises must be chosen and balanced, as usual, with the requirements coming from the mission.

References

[1] H. J. Barnaby. "Total-Ionizing-Dose Effects in Modern CMOS Technologies". Transactions on Nuclear Science 53 (2006), pp. 3103–3121.

[2] C. Redmond. "Winning the Battle Against Latch-up in CMOS Analog Switches". Analog Dialogue 35–05 (2001).

[3] R. Muller and T. Kamins. "Device Electronics for Integrated Circuits". Wiley ISBN 0-471-88758-7, pp. 441–443.

[4] C. Calligaro, V. Liberali, A. Stabile. "Design of a Rad-Hard Library of Digital Cells for Space Applications", 15th IEEE International Conference on Electronics, Circuits, and Systems (ICECS), Malta, Sep. 2008, pp. 149–152.

3

Rad-hard SRAMs

Cristiano Calligaro

RedCat Devices, Italy

> "Without change something sleeps inside us, and seldom awakens. The sleeper must awaken."
>
> Frank Herbert, Dune

In this chapter techniques for designing a rad-hard SRAM are described starting from the single memory cell (the real core of every memory design) with an evaluation of all possible trade-off between resistance against radiation and area consumption. RHBD-AL is described according to the nature of the memory (Single Port, Dual Port, Synchronous and Asynchronous) and RHBD-CL will describe main blocks providing access to the memory (DEMUX, row and columns drivers) and read/write operations. In the final part of the chapter foundations on ECC schemes will be provided.

3.1 SRAM Foundations: Single Port and Multiple Port

SRAMs are the most common devices (together with DRAMs, Dynamic Random Access Memories) used as volatile storage elements. Volatile means that information is lost when power supply is switched off. On the other hand, these so-called non-volatile memories (EPROM, E^2PROM, Flash and Emerging Memories) can store data even when the power supply is switched off.

Historically, the term "Random Access" is related to the possibility to read (or write) data directly at every address without any need to pass through other addresses or following specific paths. This was not trivial in the early stage of memory devices. Nowadays, the term RAM is usually associated with the family of volatile memory (SRAM and DRAM), while ROM (Read Only Memory) is usually associated with the family of non-volatile memory even if they have a random access approach exactly like volatile memories.

SRAM key element is of course the memory cell. As discussed in Chapter 2, the typical SRAM cell is a bistable element consisting of a latch pair (two inverters connected the output of the first with the input of the second and vice versa) with two switches gaining access to the pair. This cell, shown in Figure 3.1, is also called 6T and the inverter pair makes leverage on the CMOS technology to guarantee the stability of the state with a minimum of leakages in the junctions; a very low leakage current flows when data are stored and, indeed, under stand-by conditions, SRAM consumption is almost negligible. This cell is also known as Single Port because the access to data is guaranteed by a single pair of pass transistors.

6T SRAM cell is the evolution of the original 4T cell (see Figure 3.2) when CMOS process was not available and all transistors were n-channel only. In this case, instead of the two p-channel transistors, the load was represented simply by a resistance.

The working principle is the same, but for the transistor having the gate connected to a voltage higher with respect to its threshold, a relevant current flow produces a relevant leakage current to be provided by power supply, and producing resistors at a suitable high value was not (and it is still not) easy in the manufacturing level. Indeed, the introduction of CMOS solved many problems of power consumption and stability of written data.

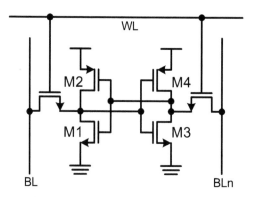

Figure 3.1   6T SRAM Cell (Single Port).

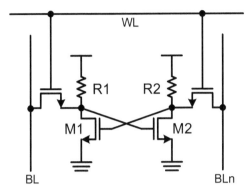

Figure 3.2 4T SRAM Cell.

Figure 3.3 shows an SRAM cell having two pairs of pass transistors gaining access to the internal bistable block. These pairs define two word-lines driving pass transistor gates and two pairs of bit-lines. This cell is known as 8T or more simply as Dual Port and represents a specific family of volatile memories that can be accessed from two directions from different digital blocks.

Figure 3.4 shows a typical example of a possible use of a Dual Port SRAM, where two CPUs have access in both read and write modes to the same memory module. In a Single Port SRAM (SPSRAM), the access should be once at a time (CPU-A first and then CPU-B), and in a Dual Port SRAM (DPSRAM), the access can be at the same time increasing the speed of internal operations. Of course, in case of an attempt to gain the same address

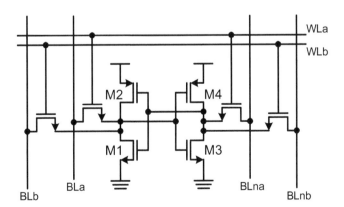

Figure 3.3 8T SRAM Cell (Dual Port).

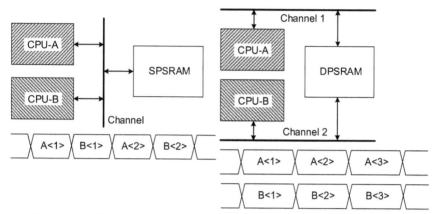

Figure 3.4 CPUs connection to Single Port (left) and Dual Port SRAMs (right).

from both CPUs, an arbitration logic must intervene to solve the issue and many methodologies are available according to the vendor solutions or the customer needs. Such arbitration systems, in a rad-hard component scenario, can be a real issue, in particular, for complex ones giving more flexibility to digital designer but, on the other hand, opening vulnerabilities related to upset (SEU, MBU and SEFI). In general, the easier is the arbitration system (at least at architectural level), the safer is the component.

Dual Port cell can be also Multi Port according to the needs of a digital ASIC. The principle is the same as shown in Figure 3.5, where a generic multiport cell is described. Usually, the number of ports can be up to three or four since the overhead coming from the decoding schemes can become too much complex and introduce problems in the arbitration system.

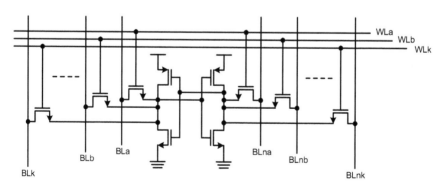

Figure 3.5 Generic Multi Port SRAM cell.

Designing a rad-hard SRAM cell is one of the most complex activities in the RHBD methodology and the main issue is related to area consumption. As described in Chapter 2, there are many solutions for making a resilient cell [1], but all of them take area and large memory cuts must consider, on the one hand, power consumption and, on the other hand, routing issues.

In principle, according to RHBD techniques, the first effort must be spent to mitigate hard errors; as is known, standard SRAMs, in particular, those coming from commercial compilers, first of all fail on latch-up. Compilers for embedded memories are not conceived to be fault tolerant, and n-channel transistors are placed at the minimum distance admitted by the lithography to p-channel transistors. Moreover, commercial SRAM cells are placed in arrays with power and ground lines running every 16, 32 or more cells; this creates distributed resistive paths that can contribute in the triggering of latch-up effects when a charged particle strikes.

Figure 3.6 shows an example of an internal array of a generic SRAM; the ground connection of the p-doped substrate is in common for up to 16 or 32 cells. Assuming that the ground line is repeated every 32 cells, the *lucky* cells will be the first and the last while the cells in the middle (like Cell 16, as shown in Figure 3.6) are the farthest to ground line. The result is that R_{sub} becomes critical in this region and the base of lateral NPN bipolar on this cell can have a weak connection to ground. In case of charge particle strike, the

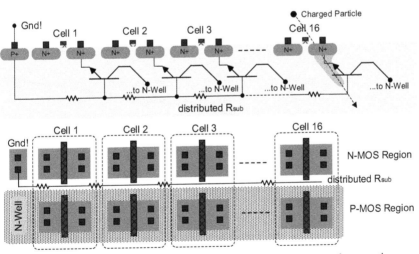

Figure 3.6 Distributed substrate resistance in a cell array having ground connection every 16 cells.

NPN can trigger the SCR (Silicon Controlled Rectifier), including the vertical PNP of p-channel transistor in Cell 16 and latch-up is activated.

The configuration of Figure 3.6 shows a series of cells becoming weaker as far as the distance from the ground line increases. In the same way, the distributed resistance in N-Well region can, in the same way, increase the probability that latch-up is triggered. Combined degradation of distributed R_{sub} and R_{N-Well} makes middle cells at high risk in hostile environments.

Latch-up of course is related to SEE, but also TID must be considered. Transistors making a 6T SRAM cell are usually at the minimum of the size offered by the technology node and the effect of borders is relevant just like in all other parts of the devices.

Making ELTs is an option, but the size of the cell increases dramatically with the drawback of a strong limitation in the final cut of the memory. Several intermediate solutions have been investigated [2], leading to different approaches suitable for specific applications and missions (LEO, GEO, HEO).

Figure 3.7 shows a possible example of a full rad-hard single port SRAM cell with 6T architecture [3, 6]. All n-channel transistors are strictly ELT while p-channel transistors are partially ELTs with a deep "U" shape. Considering that the threshold of n-channel transistors is expected to decrease with total dose, no waives have been adopted to avoid any change in the static noise margin during reading and writing. On the other hand, since p-channel transistors have an increase of threshold, a waive has been adopted with the "U" shape, thereby creating a border in the channels but reducing drastically the distance between pull-up and pull-down transistors and so creating a more compact memory cell. At very high dose (this cell has been tested up to 15 Mrad (Si)), the static noise margin makes the cell harder to be written and, consequently to be read. Pass transistors (selectors in Figure 3.7) are strictly ELT and since they have to be carefully shaped according the W/L of pull-up and pull-down they are not at the minimum of the lithography offered by the technology (a standard CMOS 180 nm in this case).

Concerning SEL, a partial enhanced guard ring has been adopted. Making a full EGR between n-channel and p-channel transistors was an excessive precaution leading to an unacceptable large cell; the choice has been to make for each 6T cell a biasing point for substrate and N-Well, as shown in Figure 3.7. In this way, there are not lucky cells like in Figure 3.6 and even if the EGR is only on one side of the pull-up and pull-down pair, the probability to switch on the parasitic SCR is at the minimum.

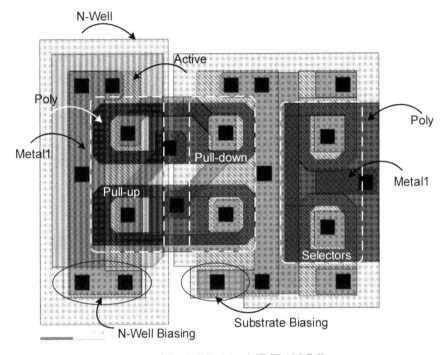

Figure 3.7   Full Rad-hard 6T SRAM Cell.

To mitigate SEUs, as mentioned in Chapter 2, a Miller Capacitor has been placed making leverage on the six metal levels offered by the technology. In this case, to reduce area consumption, a fringe capacitor approach has been chosen by using the highest metal levels (metals 4 and 5 in this case) and keeping the top metal (metal 6) for powering and grounding. Also some parts of metal 3 (not shown in Figure 3.8) contribute to the fringe, but the major part of capacitive load is offered by higher metal levels. This trick helps in increasing the critical charge of the cell and can be replicated without any interference between neighbour cells. Of course, the only limitation is that no routing can be done over the cell, thus taking to an architecture suitable for stand-alone components while for embedded memories, a different solution must be adopted, considering that in a digital ASIC, memories are usually requested to leave at least a couple of metal levels for macro-cells routing.

The array arrangement of this memory cell is shown in Figure 3.9 in a simple array of four cells. The cell is flipped along both X and Y axes to guarantee N-Well regions (with p-channel transistors) and to create a common polysilicon region for selectors of contiguous cells along word-lines

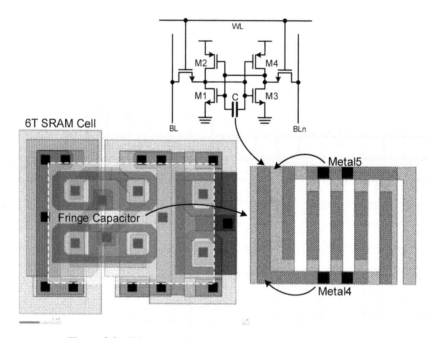

Figure 3.8   Fringe capacitor on 6T SRAM cell to mitigate SEUs.

(WLs). In this way, a better powering strategy is guaranteed, thus reducing at the minimum any possibility to trigger parasitic SCRs.

The array arrangement itself is the sum of the three RHBD approaches described in Chapter 2 (architecture, circuit and layout).

The architectural approach is in the organization of the cell to guarantee an array "easy" to be powered in the proper way (mitigation of SELs); the circuit approach is in the cell (e.g. Miller Capacitor) to mitigate SEUs; the layout approach is in the ELT design and in the introduction of biasing (both for power and ground) in the right position.

Figure 3.10 shows the organization of the array in a 16 bit-lines to be compared with the same array of Figure 3.6 of a standard SRAM (not rad-hard). These are the regions where an SEL may occur getting in connection N-Well with array p-substrate.

The specific architecture of this 6T SRAM cell makes it possible to have power and ground "islands" inside the array facing each other. This island, shaped as a sequence of "ground/power/ground", represents a small EGR in the middle of four SRAM cells, and parasitic SCR in all directions have a cut in the positive feedback loop on the distributed resistance both coming from

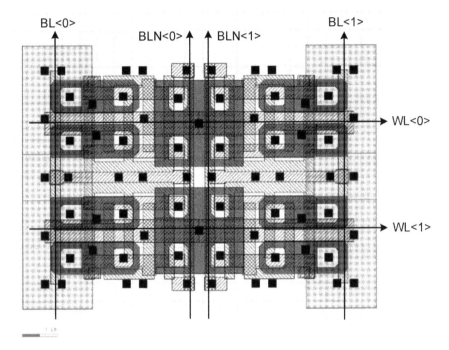

Figure 3.9 Array of four 6T SRAM cells organized in word-lines (WLs) and bit-lines (BLs).

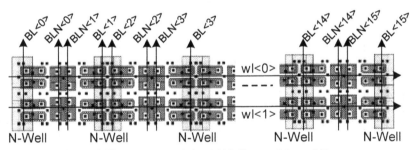

Figure 3.10 Array containing 16 bit-lines and 2 word-lines.

the substrate and from N-Well region. This RHBD technique is called IGR (Island Guard Ring), and islands can be connected, according to Figure 3.11, with parallel metal lines running on higher metal levels (e.g. metal 3 or metal 4).

A good power and ground strategy is always beneficial also in standard memories: in rad-hard, it becomes crucial. Usually, this can be made with very different strategies, but in general, all of them are focussed in making

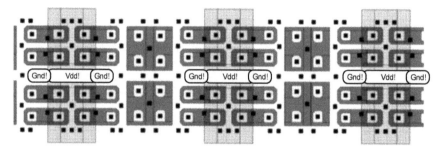

Figure 3.11   Island Guard Ring (IGR) strategy in a 16 bit-lines and 2 word-lines array.

the array independent from the rest of the periphery and, if possible, with an its own ground and power pad. Of course, this is not always possible, and in embedded memories, this can be a real issue for digital designers summoned to implement a macro expected to be Plug&Play.

In stand-alone memories, where a higher freedom degree is available, a possible powering strategy can consider high metal levels for making running power supply and ground. Figure 3.12 shows a possible strategy,

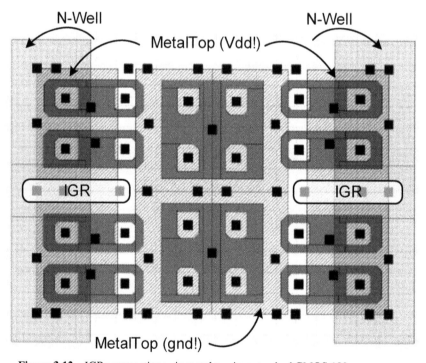

Figure 3.12   IGRs connections via metal top in a standard CMOS 180 nm process.

where the top metal (metal 6 in this specific example) is the one connecting IGRs in the array. Usually, silicon foundries offer two types of top metals: thin and thick. Some of them also make copper available for the top metal to be able to provide a low resistance and it is very suitable for powering issues.

In Figure 3.12, only N-Well, poly, contacts and top metals are shown, and it is possible to see how top metals have been designed to be as large as possible to make power and ground running in parallel at the minimum of the distance admitted by the lithography in the process. This follows the philosophy of the guard ring, and if the top metal is copper, the result is an array with a very stable powering and grounding not leaving any space for distributed resistances or voltage drops. The overall result is an array having a very low probability to trigger any SEL and, thanks to ELTs, to have a progressive degradation both in the intra-device directions (the channel of the transistor) and in the inter-device leakage (between different transistors). It is worth to underline that degraded transistor from TID can open weaknesses in terms of parasitic SCRs. In fact, inter-device leakage between n-channel and p-channel transistors can contribute to biasing of the base of the lateral NPN transistor represented by the substrate of the CMOS technology.

The cross correlation between TID and SEL (degradation due to TID causing SEL) is the classical combined effect of two different known problems in radiation hardening, and it is a matter of recent research from space agencies.

Making such a strong 6T SRAM cell like the one shown in Figure 3.7 is not always possible. The major limitation is the area consumption, but indeed not all missions need an SRAM with a TID over 15 Mrad (Si). LEO missions, according also to the envisaged time frame, can require 100 krad (Si) or less, so it makes no sense designing such a rad-hard memory.

Indeed, RHBD techniques should consider "how much" rad-hard must be the device and then apply the right methodology for the right mission also considering the suitable silicon process.

RHBD is the arena of compromises. The number of freedom degrees is high and everything should start from mission requirements on one side (which application, which orbit, how long is the mission) and from the selected process on the other (which node, which oxides, how oxides are built). Then, an RHBD roadmap can be defined pushing more on some aspects and relaxing on others to gain what usually customers ask for: storage density, high speed, limited number of MBUs and, last but not least, area consumption.

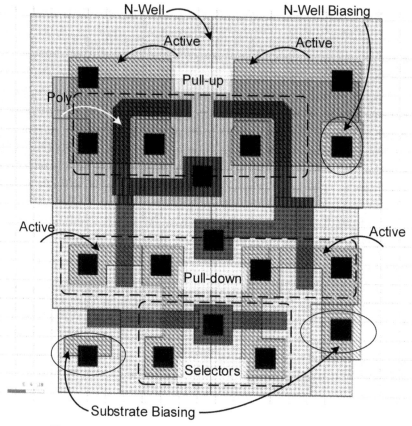

Figure 3.13   Rad-hard 6T SRAM cell for embedded applications.

Figure 3.13 shows an example of a rad-hard 6T SRAM cell conceived for embedded applications, where several waivers have been adopted to minimize area and power consumption. The first, and most evident, waiver is regular shape of both n-channel and p-channel transistors instead of the canonical ELT shape. The technological process behind is a standard 180 nm CMOS with dry oxidation that can guarantee a good level of stability and a low number of traps in STI regions.

The adoption of a regular shape in transistor channels with dry oxidation has been proven up to 300 krad (Si), which is still much lower to the performance of 6T SRAM cell of Figure 3.7 (15 Mrad (Si)) but enough for the major part of LEO and GEO missions.

Concerning SEL, to minimize area, neither EGR nor IGR could be adopted, but three biasing points have been placed: one for N-Well, one and a half for ground. The half-substrate biasing contact on the lower left side of the cell is shared with another neighbour cell. This approach privileges substrate biasing with respect to N-Well, and it is because the main cause of parasitic SCR activation is the distributed resistance on p-substrate to open the possibility of a voltage spike (from a charged particle strike) driving the base of the lateral NPN bipolar. Having the possibility to choose, it is always better to sacrifice N-Well biasing with respect to p-substrate biasing.

Even if both EGR and IGR have been excluded, nevertheless the RHBD strategy for SEL adopted in Figure 3.13 has been demonstrated to be enough up to 60 MeV*cm^2/mg by using Xe ions [4, 5].

Concerning soft errors (SEUs), the approach is very similar to the one adopted in Figure 3.8.

In this case, the Miller Capacitor (see Figure 3.14) is smaller than the previous one and considering that the critical charge of the bistable pair is lower because of reduced W/L of both n-channel and p-channel transistors (compared to ELTs counterparts), the overall result is that this cell is more prone to soft errors than the one in Figure 3.8. Moreover, as mentioned previously, it can happen that not all metal levels can be available (e.g. process with only four metal levels like those used in image sensors) and in this case, there is no possibility to build the Miller Capacitor at all and the 6T

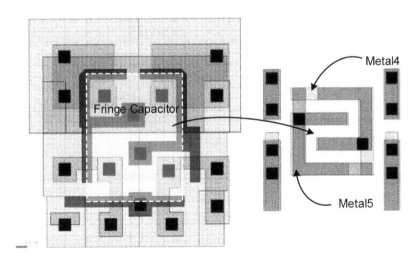

Figure 3.14   Reduced fringe capacitor on 6T SRAM cell to mitigate SEUs.

SRAM cell can count only on the capacitance offered by polysilicon gates of the bistable pair. In this case, the only RHBD countermeasure that can be adopted is to make poly-lines as large as possible and occupy empty areas (if available) with polysilicon connected input nodes of pull-up and pull-down pairs.

Figure 3.15 shows a comparison of the two different approaches described above. The first cell (left) is almost 13 μ^2, while the second (right) is 8 μ^2. This difference becomes relevant both in a stand-alone approach and in particular in an embedded approach. A 4Mbit device, just to make and example, only from the array point of view takes around 55 mm^2, while the second approach is around 34 mm^2. Considering that rad-hard periphery takes area, it is clear that pushing a chip to 70 or 80 mm^2 can create yield and power issues.

As known to memory designers, making memories too much extended is possible (and indeed mature technologies still maintain a good yield level also with large areas) but the drawback is to have power and consequently thermal issues. This is particular true for synchronous memories. Power and

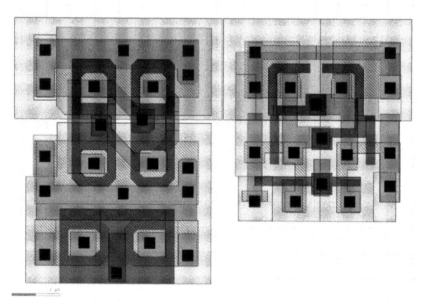

Figure 3.15   Size comparison between 6T SRAM cells adopting different RHBD techniques.

ground strategies are crucial in a rad-hard design (we discussed extensively in this paragraph) and making large chips requires a further effort that can become a problem in the overall economy of a rad-hard component. Making more supply and ground pins is of course an option and making a physical separation between matrix array and peripheral circuits is beneficial as well; but if a single chip memory reaches a size of, for example, 100 mm^2, then this can be too much and power budget can be unacceptable by end users (not considering packaging issues).

For embedded memories, the situation is even more critical. A full rad-hard 6T SRAM cell can be very hardly accepted as an L1 Cache (or it must be a very specific application) while the second one is more suitable. In this case, according to end user requirements, it can be implemented with Miller Capacitor or not. Also in this case, it depends on the ASIC design: if an ECC is envisaged (this happens very frequently), maybe it is better to avoid the Miller Capacitor and make a stronger ECC algorithm; alternatively, the Miller Capacitor may be introduced (so rising the critical charge and consequently the LET SEU_{th}) to relax the ECC strategy with a more compact algorithm. RHBD efforts do not finish when a component or a macro is delivered to an end user but continue under different premises and trying to get the best from already designed rad-hard building blocks to be implemented in a larger design.

All considerations done with Single Port SRAM cells are valid also for Dual Port cells. Figure 3.16 shows a typical example of 8T cell designed without ELT and with n-channel transistors almost at the minimum offered by the technological process (CMOS 180 nm). This cell is designed by using only three metal levels, thus making it suitable for embedded applications requiring a routing over macro approach.

N-Well and substrate biasing still have neither EGR nor IGR, but they have been placed on the same side of the cell at least to optimize power and ground routing and to make more symmetric the memory cell. N-Well and substrate biasing are connected to a neighbour cell flipped on Y axis to build a single region between two cells where to cut positive feedback loop of parasitic SCR.

Just like the previous cells, this Dual Port can be completed with a Miller Capacitor to increase the critical charge of the bistable pair or, if higher metal levels are not available, some additional poly regions may be added to gain capacitance on the inputs of the bistable pair.

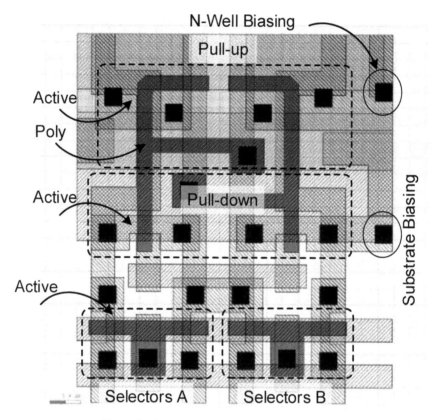

Figure 3.16   Rad-hard 8T (Dual Port) SRAM cell.

3.2 Synchronous or Asynchronous?

SRAMs can be synchronous or asynchronous according to the application and, in principle, it does not matter if they are stand-alone or embedded devices. In a hostile environment like space, this discrimination is relevant; synchronous memories are more prone to soft errors because of the path represented by the clock tree(s). As described in Chapter 2, transients (SETs) may use the clock tree for an internal propagation and once reached critical blocks, such flip-flops or state machine, they can trigger upsets or functional interrupts.

In SRAMs, synchronous and asynchronous behaviour is described by Figure 3.17. The main difference is that in the first case, a clock signal is provided to the whole memory and all operations are strictly driven by it

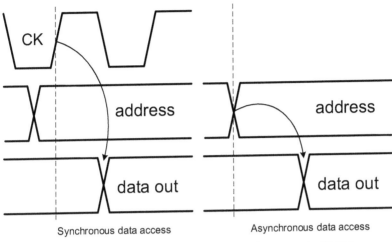

Synchronous data access   Asynchronous data access

Figure 3.17   Synchronous and asynchronous data access in SRAMs.

rising (or falling) edge. If a data byte is going to be read, it is necessary that address bus and control signal are already stable before the rising edge of the clock; the access time is the time interval between the rising edge of the clock and the data byte available. On the other hand, asynchronous memories have any clock signal and timing is generated by internal circuits called Address Transition Detection (ATD), providing internal timing for making data available. In this case, the access time is the time interval between the transition of the address (or other control signal) and the effective available data out.

In synchronous SRAMs, everything run around the clock, which is the master of all operations, and since the other signal such as Chip Enable (CE), Output Enable (OE), Write Enable (WE) and address are provided externally, it is necessary that they are applied in the right way to avoid any interference with clock. In Figure 3.18, time parameters are shown in this case only for address signals (but for CE, OE and WE, the same parameters applies). If an address has to be provided, it must be stable before the rising edge of the clock for a minimum interval time called Setup Time, and once the rising edge of the clock is over (thus triggering all internal operations for reading or writing), address must be kept for an interval time called Hold Time.

Setup Time and Hold Time are usually specified at datasheet level and guarantee the right working condition for a memory giving the specified access time. Setup Time and Hold Time also pave the way for the cycle time of the memory, which is the interval between two rising times of the clock.

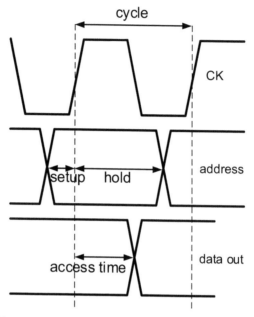

Figure 3.18   Time parameters in synchronous SRAMs.

In asynchronous memories, since everything is triggered by a control signal or an address, there are still such requirements but are related to the signal triggering the read/write operation; of course, this causes some complexities in the circuitry of the memory. In general, asynchronous memories are more robust than synchronous memories under irradiation and this happens because there are any paths for make running transient in the devices.

As shown in Figure 3.19, the clock signal in a synchronous memory drives main blocks such as input buffers, DEMUX blocks for accessing word-line and bit-lines, the Read and Write circuits and Equalizers for taking bit-lines under the best conditions before reading and writing. The only block that in principle is not reached by clock signal is the matrix array which is the real content of data. An SET occurring on the input of one of these blocks in the clock distribution can be propagated inside all these blocks. Drivers are used to give strength to clock signal and such drivers can be the weak point of the whole.

Asynchronous memories, on the other hand, do not show this problem and ATD circuits are usually distributed inside the chip and very specific for

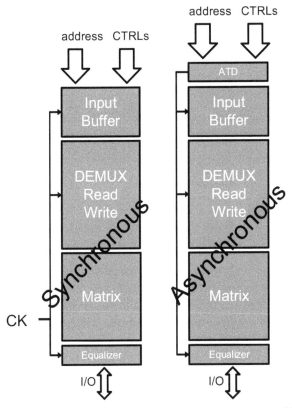

Figure 3.19 Basic architecture of synchronous and asynchronous SRAMs.

each block, thus preventing any contamination when an SET occurs. It is this intrinsic characteristic making asynchronous SRAM more reliable in terms of soft errors. Moreover, ATD engine is activated only when a transition is detected (and it is not always active like clock) so the time frame where a charged particle can hit provoking an SET is very limited, thus reducing the probability of occurrence of such an event. Apart from these assumptions, making rad-hard synchronous SRAMs is possible, only particular care must be taken at architectural level design. Clock gates, extensively used in large digital ASICs to save power, can be beneficial, and the adoption of rad-hard digital cells stopping SET propagation is mandatory.

Since the triggering signal can vary in asynchronous SRAMs, parameters must be specified accordingly. Figure 3.20 describes the working condition of an SRAM activated by CE (or CE-driven). In this case, the other signal must

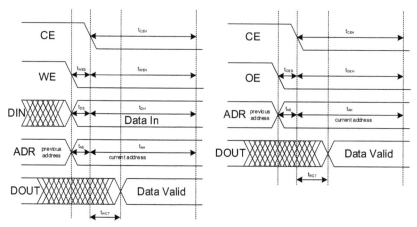

Figure 3.20   CE-driven Write Mode (left) and Read Mode (right) in an asynchronous SRAM.

obey to a Setup Time with respect to CE and have a Hold Time again with respect to CE. This is valid both for write mode and read mode, which means taking down WE before CE falling and, on the other hand, taking down OE before CE falling. Of course, during write mode, OE must be kept high and during read mode, WE must be kept high.

If CE is kept low in both write and read modes, it possible to have a write operation mastered by WE (WE-driven) or a read operation mastered by OE (OE-driven). Figure 3.21 shows Setup and Hold times for all the other signals.

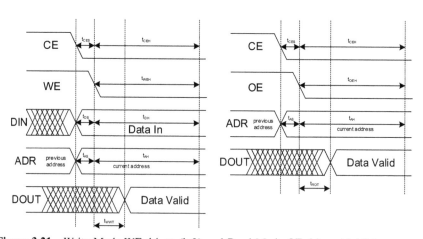

Figure 3.21   Write Mode WE-driven (left) and Read Mode OE-driven (right) in an asynchronous SRAM.

The last, and indeed the most common, write and read mode is the address-driven one, as shown in Figure 3.22.

In synchronous SRAM, both read and write modes are mastered by clock signal and all Setup and Hold times are related to it as shown in Figure 3.23.

Setup and Hold times are used in datasheets (see Figure 3.24) to guide end users in the application on-board if the SRAM is a stand-alone component,

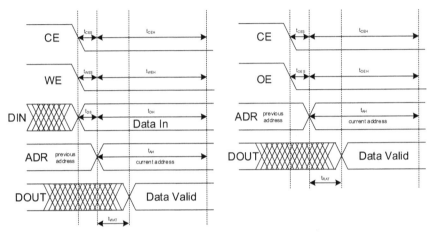

Figure 3.22 Write Mode (left) and Read Mode (right) both address-driven in an asynchronous SRAM.

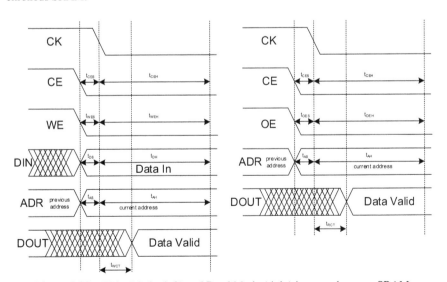

Figure 3.23 Write Mode (left) and Read Mode (right) in a synchronous SRAM.

```
36 specify
37   specparam
38       DF = 1.0,
39       tACC =    *DF,        // access time
40       tAS  =    DF,         // Address setup time
41       tAH  =    DF,         // Address hold time
42       tDS  =    DF,         // DIN setup time
43       tDH  =    DF,         // DIN hold time
44       tWES =    *DF,        // WE setup time
45       tWEH =    *DF,        // WE hold time
46       tOES =    *DF,        // OE setup time
47       tOEH =    *DF,        // OE hold time
48       tCES =    *DF,        // CE setup time
49       tCEH =    *DF,        // CE hold time
50       tCKS =    *DF,        // CK setup time
51       tCKH =    *DF,        // CK hold time
52       tWCE =    F,          // CE Width
53       tWOE =    F,          // OE Width
54       tWWE =    F,          // WE Width
55       tWA  =    ,           // A Width
56       tWCK =    F;          // CK Width
57
58   $setup(posedge A,posedge CK, tAS); $setup(negedge A,posedge CK, tAS);
59   $setup(posedge DIN,posedge CK, tCKS); $setup(negedge DIN,posedge CK, tCKS);
60   $setup(negedge WE,posedge CK, tWES);
61   $setup(negedge OE,posedge CK, tOES);
62   $setup(negedge CE,posedge CK, tCES);
63
64   $hold(posedge CK, A, tAH);
65   $hold(posedge CK, DIN, tCKH);
66   $hold(posedge CK, WE, tWEH);
67   $hold(posedge CK, OE, tOEH);
68   $hold(posedge CK, CE, tCEH);
69
```

Figure 3.24   Excerpt of a Verilog model file of a synchronous Single Port SRAM.

or can be implemented in a model (Verilog or VHDL) if the device must be embedded in a larger ASIC. In this case, Setup, Hold, Cycle and access times become critical parameters in digital simulations.

3.3 SRAM Architectures

As described in the previous paragraph, SRAMs may be asynchronous or synchronous and because of this nature, they usually have the signals shown in Figure 3.25.

Chip Enable (CE or sometimes CEN) is the signal that activates the SRAM or keeps it in stand-by. In general, this signal is used to gain access to a specific memory block having previously stored data; when neither reading nor writing are performed, CE keeps the memory in stand-by to limit power

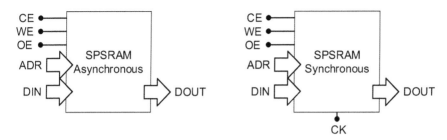

Figure 3.25 Single Port SRAMs (SPSRAMs).

consumption. Write Enable (WE or WEN) is the signal adopted for taking the memory in write mode, excluding the output data bus (DOUT) and enabling the data in bus (DIN) to give data to cell matrix array. Output Enable (OE or OEN) is the pin enabling the read mode condition of the memory enabling the data out bus (DOUT) and inhibiting the data input bus (DIN). ADR is the address bus, which is considered during both reading and writing. Synchronous SPSRAMs have an additional pin represented, as described in the previous paragraph, by the clock (CK).

In addition to the main functions described above, it is possible to have other sub-functions according to the vendor or coming from specific applications; just to give an example, one of this is the read-through option coming when DIN bus is acquired during a write operation on a specific address; in this case, DOUT bus is not inhibited but gives out the same data written in the memory array.

In synchronous SPSRAMs, the clock drives all operation and all pins and data must have their Setup and Hold Times including DIN bus. On the other hand, asynchronous SPSRAM do not have such requirements except DIN bus, which is the only one not triggering any operation. Conventionally, the signals detected by ATD circuits are CE, WE, OE and ADR.

Dual Port SRAM (DPSRAM) are like two memories sharing the same matrix array. There are two sets of control signals (CE, WE and OE) and two set of addresses (see Figure 3.26); also input and output are independent because the principle of DPSRAM is that they can be accessed by two external devices simultaneously. DPSRAMs can be asynchronous and synchronous as well but in the latter case, the clock signal is only one and all setup and hold times are related to it; somewhat more complex is in asynchronous DPSRAM where two independent ATD systems play the role of triggering write and read operations according to a number of setup/hold times coming from two different sides.

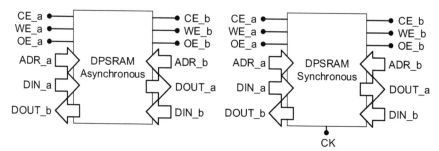

Figure 3.26   Dual Port SRAMs (DPSRAMs).

Figure 3.27 shows the main blocks comprising a typical SPSRAM. Both synchronous and asynchronous share the major part of these blocks and only the latter has a dedicated block (ATD). From the matrix point of view, it

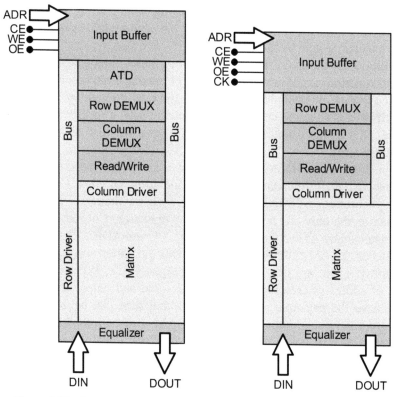

Figure 3.27   Basic architecture of synchronous and asynchronous SPSRAMs.

is worth to underline that row and column drivers are usually considered as part of the array and, as described in the previous chapter, in a rad-hard memory, they play a significant role in making a resilient (and efficient) access to word-lines and bit-lines. Usually, terminal drivers are, in standard memories, designed to be small (to maintain the pitch of the memory) and can drive more memory banks (more bits); in the case of rad-hard memories, they must follow all possible RHBD rules to avoid MBUs or provoke SETs during reading and writing.

Read/Write block can be quite different in synchronous or asynchronous SPSRAM, but in principle, it is based on the sense that amplifiers drive the first directly by the clock and the latter by ATD circuitry. This block is critical from radiation point of view and requires a very careful RHBD approach both at circuit and at layout level (see Chapter 2).

Row and Column DEMUXs are practically the same and can be used transparently in synchronous and asynchronous devices since they are mainly digital blocks where the use of a robust rad-hard digital library can solve the major part of the problems; the only difference can be the role of CK that in the asynchronous counterpart becomes an internal signal coming from ATD block.

Input buffers for both control signals and address play the role of strengthening the signals running in the core of memory periphery; usually, they also provide the negative control and address to make DEMUX operations easier and more efficient. In addition to canonical rad-hard buffers (that can come from a digital library), usually delay elements are utilized in synchronous SRAMs to perform disoverlapped operation, in particular in sensing scheme (Read Mode), and in asynchronous SRAMs to generate internal phases driving decoding and sensing/writing schemes. Input buffer block may be the source of cascade errors inside the peripheral blocks of the memory and, for this reason, RHBD techniques must be severe, in particular with delay elements. The latter elements should include capacitors built using polysilicon and having in input and output buffers to avoid any propagation of SETs. A wise strategy in input buffers and delay elements is mandatory to have efficient DEMUXs for addressing in the proper way word-lines and bit-lines.

Figure 3.28 shows the Dual Port counterpart of Figure 3.27, in which it is possible to see how the same considerations done for Single Port can find an application for DPSRAMs. In this case, the only difference is that peripheral blocks are replicated and accessed by pairs of addresses and control signals.

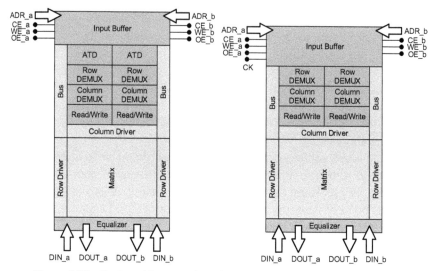

Figure 3.28   Basic architecture of synchronous and asynchronous DPSRAMs.

Making an architecture strong enough to resist to soft errors is a matter of isolating those parts known to be weak (e.g. row and column drivers) and mitigating their effects in case of charge particle strike.

Figure 3.29 shows a possible architecture of a generic 8-bit SPSRAM, where the matrix is divided into eight independent arrays each with its final stages. In this case, if a charged particle hits a column terminal driver thus enabling the wrong bit-line, the result is that a single error will occur since the other terminal drivers are not affected by the strike.

If the charged particle strikes, for example, a Row DEMUX in this case, this block, in common for all the bits, can really decode a wrong word-line independently from the terminal driver and, in this case, the transient becomes an MBU (all the bits are wrong).

The approach of dividing and making independent critical blocks may be applied to a complete memory (i.e. eight independent memories), but of course this is risky because area consumption can get out of control. From a sustainable rad-hard architecture, it is necessary to go down to a strong enough digital block. In the example of Figure 3.29, DEMUXs can be designed with some internal redundancies like the one suggested in Figure 2.26 possibly in the pre-DEMUX part; in this case, a simple voter can take in safe condition any possible transient hitting directly the pre-decoder or coming as an SET from the Input Buffer block.

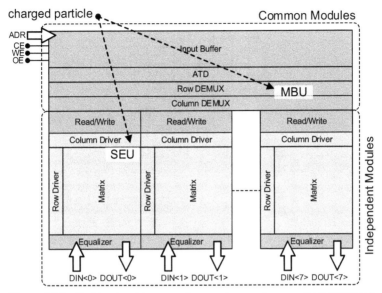

Figure 3.29 Rad-hard architecture of a generic asynchronous 8-bit SPSRAM.

The massive replication of terminal drivers is very frequent in rad-hard memory, since this circuit is usually small and affected by severe limitation in layout design; making it really rad-hard (at least from the soft error point of view) is an issue and, in principle, the design is focussed on making it resilient from TID and SEL (see Figure 3.30); placing countermeasures for soft errors (e.g. additional capacitance) is very complex and risky to violate the pitch.

For the independence of DEMUXs, it is also possible to choose a mixed approach like the one suggested in Figure 3.31. In this case, the row DEMUX is the same for all the memory, while column DEMUX is replicated for every bit; this approach can be valid for a memory array having a high number of rows and a low number of columns. In this case, the column DEMUX is expected to be compact and the replication can affect in a limited way the whole design.

According to Figure 3.31, if a charged particle hits the DEMUX layer of the memory, the result is an SEU instead of an MBU; from another point of view, a replicated column DEMUX able to give only SEUs can be relaxed in RHBD strategies, thus reducing the impact on the device in terms of area and power consumption. As usual, in RHBD, it is a matter of balancing replication of light structures or making one structure only with a high level of precautions.

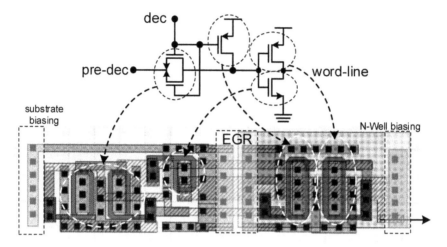

Figure 3.30   Rad-hard word-line terminal driver.

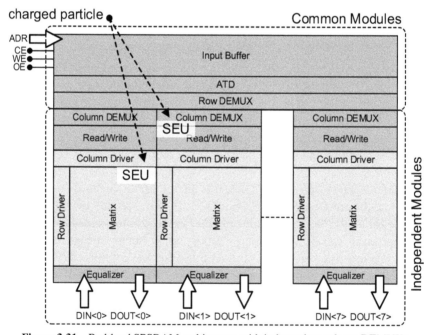

Figure 3.31   Rad-hard SPSRAM architecture with independent column DEMUXs.

Choosing how to handle from the architectural point of view DEMUXs is strongly related, as mentioned above, to array size. In stand-alone memories, where it is supposed that the minimum cut of an SRAM is 256 or 512 kbit, it is evident that a simple replication of DEMUXs is not viable.

Figure 3.32 shows the top-level floor-plan of an asynchronous 512 kbit rad-hard SPSRAM organized as 64 kbit × 8. The eight bits are organized with separated arrays each having independent terminal drivers; four blocks are faced in the lower part of the chip while other four in the upper part. In the middle, there are the common blocks including DEMUXs, ATD and Input Buffer.

The same chip organization is also evident in the photo shown in Figure 3.33, in which a commercial rad-hard SRAM (RC7C512RHM) is assembled in a standard ceramic package. It is possible to see in the photo

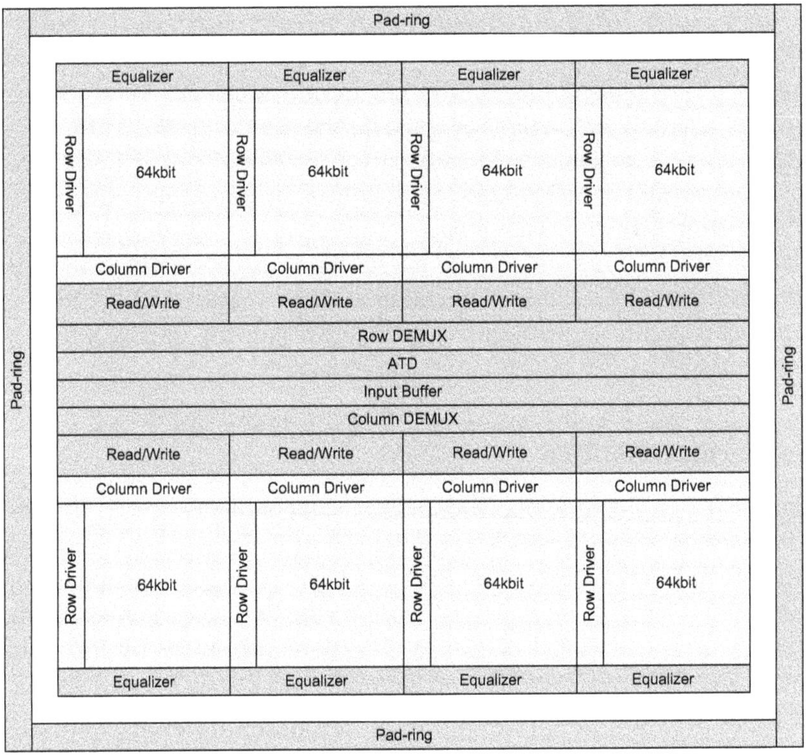

Figure 3.32 Rad-hard top-level floor-plan of a generic 512 kbit (64 kbit × 8) SPSRAM.

Figure 3.33   Chip photo of RC7C512RHM implementing the architecture of Figure 3.32.

how memory blocks (64 kbit) are separated from each other; between memory blocks, there are common blocks, as explained above. The chip has been integrated with a standard CMOS 180 nm from TowerJazz, where, at the process level, the only precautions adopted has been a dry oxidation to make STI cleaner than in a standard oxidation. RC7C512RHM has demonstrated to be resilient to TID over 300 krad (Si) with Cobalt 60 and SEL-free up to 60 MeV*cm^2/mg (Si) with xenon ions.

Concerning soft errors, this SRAM has used a 6T SRAM cell like the one described in Figure 3.13 and implementing a Miller Capacitor; LET SEU$_{th}$ has been calculated to be between 4 and 5 MeV*cm^2/mg (below nitrogen ions) with only SEUs. No MBUs have been observed.

3.4 Embedded SRAMs

In digital ASICs (e.g. DSPs, Microcontrollers and core processors), SRAMs play a relevant role and usually a large portion of a digital chip is covered by memories. In core processors, an SRAM (it can be either synchronous or asynchronous) is usually located very close to the core and running at the same frequency: this is called L1 Cache [7]. The requirements for this kind of memory are very high (starting from the frequency demand) and their role is crucial in obtaining a good performance of the overall digital device.

In space applications, frequencies are usually much more relaxed, but of course, the main issue is to guarantee a rad-hard device avoiding an excessive use of silicon area limiting the overall performance.

Rad-hard Embedded SRAMs must fulfil requirements somehow different from the stand-alone ones; oxide degradation and latch-up issues must follow the same approach, but for soft errors, different strategies must be considered. Embedded SRAMs usually do not have the same size of stand-alone and the organization is different since the word can be 16 or 32 bits with a depth varying from 1 kbit to 8 kbit. Since soft errors are usually handled with ECC engines, different algorithms can be applied according to the needs of the application.

Under these premises, SRAMs must be scalable and inclusive of additional bits required for ECC algorithms (e.g. 16 bits for data storage and 4 bits for ECC) but without any logic, since end user must have the freedom to choose where to store data and to synthesize its own ECC engine.

Modularity is the key for making rad-hard embeddable SRAMs, and Figure 3.34 shows an example of blocks replicated n-times and sharing on top the same address bus and signal controls. Just placing SRAM blocks as an array, the memory is built with a minor effort from the designer of the compiling engine.

In this case, everything is replicated, including row and column DEMUXs; this can be convenient for cuts between 2 kbit and 8 kbit, and of course, the result is a memory module affected only by SEUs (MBUs cannot occur because each bit is completely independent), but of course, there is a penalty in terms of area.

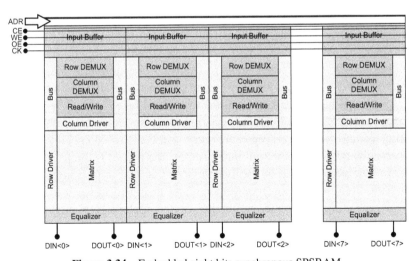

Figure 3.34 Embedded eight bits synchronous SPSRAM.

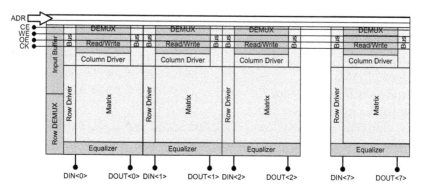

Figure 3.35   Embedded eight bits synchronous SPSRAM with common Row DEMUX and Input Buffer.

Figure 3.35 shows a different solution where Row DEMUX and Input Buffer are in common for all the memory modules. From the area point of view, there is a clear advantage (even if common blocks must be sized to be able to drive not one but eight blocks), while from soft error point of view, there is the possibility that a charged particle strikes common input buffers or the DEMUX, thus causing an MBU.

From the digital application point of view, the solution of Figure 3.34 requires only an SEC (Single Error Correction) ECC, while the solution of Figure 3.35 needs at least an SECDED (Single Error Correction Double Error Detection) ECC. However, to implement SEC, four additional bits are enough, while to implement SECDED, at least five bits are required. This means that the latter needs an additional SRAM module that can take part of the area saved by using common buffers and row DEMUX.

Making a trade-off between area (and power) consumption, mitigation of soft errors and ECC algorithm complexity is part of RHBD efforts; just like making a rad-hard 6T SRAM cell fully shielded (see Figure 3.7) makes non-sense for an LEO application, also providing an excessive soft-errors mitigating system, which can represent a non-sense for some specific applications. As always happens in rad-hard systems, the higher is the protection, the lower is the performance of the final device so it is of utmost importance knowing the final application requirements and trying to target it avoiding excessive precautions or under-estimation of the real performance of the system.

Figure 3.36 shows an example of a synchronous SRAM organized as 8 kbit × 8 (64 kbit) and generated by a compiler. In this case, the architecture chosen is fully replicated so that only SEUs are expected and, for this reason, only four additional modules (plus ECC algorithm) are necessary to have a fully soft-error mitigated device.

Address and control signals run in the upper part of the memory on a common bus rail and other modules can be added extending the rail to feed all memory banks; in the lower part, there are data in and data out.

According to Figure 3.36, all memory banks are placed in line, but different shapes may be considered. In Figure 3.37, the eight bits are organized four on one side and the four on the opposite side. In the middle, there are input buffers with rail carrying address and control signals running exactly over the buffers to minimize the resistive path along metal lines.

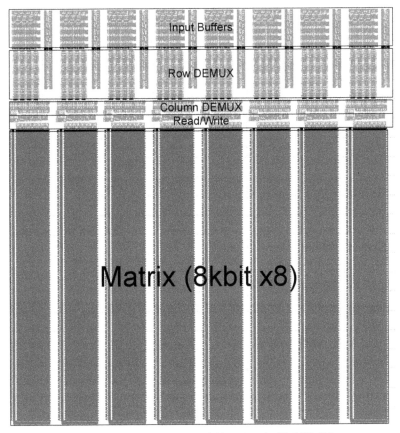

Figure 3.36 Synchronous 8 kbit × 8 embedded SRAM.

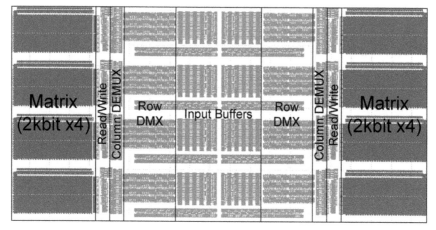

Figure 3.37   Synchronous 2 kbit × 8 embedded SRAM.

3.5 SRAMs' Building Blocks...Rad-hard of Course

It has been always difficult to categorize memories as analog or digital and even if in the end they are considered digital in their core, many parts can be considered, and indeed designers deal as, analog. Sensing schemes used for every kind of memory (SRAMs, Flash, etc.) are based on amplifiers or comparators and the way how bit-lines are driven to obtain the best conditions to make sense working (equalization and precharge) follow design rules much closer to analog design than digital.

But some other parts are strictly digital; all DEMUXs are based on digital cells and sometimes these parts are built following a digital flow including synthesis and automatic Place & Route.

3.5.1 Input Buffers and ATDs

The role of input buffers is to take signal coming from the external and distribute them into the SRAM also in the negate form; control signals are CE, WE and OE, while address are represented (see Figure 3.38) by the bus ADR<0:n>, where n is the most significant bit of the memory address. If the SRAM is asynchronous, then there is also ATD (Address Transition Detection) that usually is embedded with input buffers (for layout reasons). If n is 10, then the SRAM has a depth of 1 kbit, 2 kbit of 11, 4 kbit for 12 and so on.

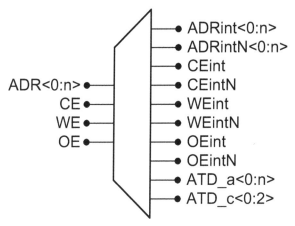

Figure 3.38 Input Buffers and ATDs in an asynchronous generic SRAM.

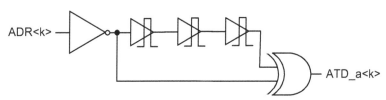

Figure 3.39 ATD generator at k address in an asynchronous generic SRAM.

Each address and each control signal (CE, WE and OE) must generate a transition if the device is asynchronous and this is obtained by a very simple circuit, as shown in Figure 3.39.

The goal is to generate a single pulse when the address has a transition from high to low or vice versa, which is achieved by using a series of delay elements. The input signal (ADR<k>), after a first negation having the purpose mainly to square and give strength, is connected directly to one input of XOR port, while a delayed copy of the same signal is taken to the second input of the XOR. Since the XOR gives a "0" when its inputs are equal (both "0" and "1"), the "1" will have a length equal to the sum of delay elements taking ADR<k>.

Figure 3.40 shows a possible layout of the scheme of Figure 3.39, where it is possible to see how delay elements have been designed as a series of an inverter, a capacitor (represented by large polysilicon area in standard n-channel and p-channel transistors) and a final inverter to straighten the input signal. This solution has the advantage to filter SETs if a charged particle hits the capacitor and, thanks to the possibility to modulate the width of the

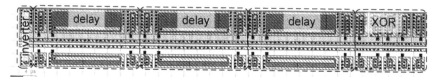

Figure 3.40   Physical layout of an ATD generator in a standard CMOS 180 nm process.

transistors, to make a fine tuning of the capacitance thus obtaining the right delay with a good level of stability in temperature.

Since control signals in an asynchronous SRAM are always evaluated in the negate form (SRAM active when CE is "0", Write Mode when WE is "0" and Read Mode when OE is "0"), the ATD in this case has to consider only the passage from "1" to "0" and not vice versa. This is obtained with a change in the scheme of Figure 3.39 simply filtering the transition from low to high with an AND port in cascade to the XOR and having at the second input the negate of the original signal to be detected.

Figure 3.41 shows the physical layout of a buffer, which is basically composed of an inverter trumpet with a single inverter as a first stage and four

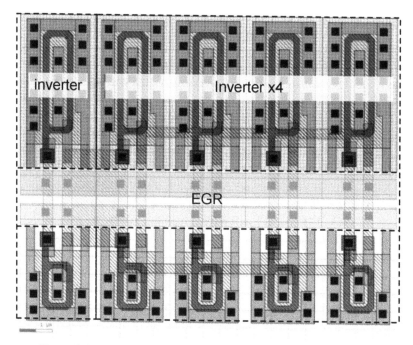

Figure 3.41   Physical layout of a buffer as a trumpet of inverters 1 × 4.

inverters in parallel giving strength to the output signal. It is possible to see how the major part of RHBD rules are strictly followed to avoid TID (ELTs), SEL (EGR) and SET propagation. For buffers requiring a higher strength, a possible solution can be represented by a trumpet having two inverters in parallel as a first stage and eight inverters in parallel as a second stage with a physical design very similar to the one of Figure 3.41.

3.5.2 DEMUXs

DEMUXs are the digital part of a generic memory design, and, in this case, having a good rad-hard library can be an option for making blocks reliable and spending a minor effort in the layout design by using automatic Place & Route tools.

Figure 3.42 shows one of the main building blocks of a generic DEMUX (valid for both rows and columns) built by an AND port (with four inputs), a NAND handling an enable signal (EN) used to keep in stand-by the DEMUX and providing phases coming from ATD engine and a buffer giving strength to the output signal expected to drive terminal stages directly facing to word-lines and bit-lines. Exactly like the previous block (Figure 3.42), EGR and ELTs guarantee immunity to hard errors coming from SEL and TID.

The block shown in Figure 3.42 is repeated to fit all possible combinations of the four inputs AND (2^4 outputs), thus leading to 16 basic DEMUX organized as shown in Figure 3.43, where only the first two are represented. To obtain homogeneous regions in N-Well and p-substrate, the cell of

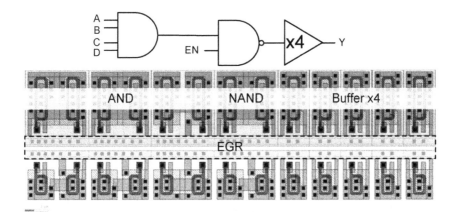

Figure 3.42 Physical layout of a basic 4 bits DEMUX.

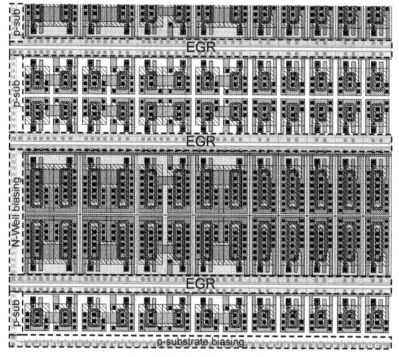

Figure 3.43   DEMUX ensemble.

Figure 3.42 is flipped on odd and even rows; in this way, also thanks to end-caps at the beginning and at the end of the row containing the basic DEMUX, it is possible to close all N-Well regions (containing p-channel transistors) and p-substrate regions (containing n-channel transistors) separating each other and so fulfilling all RHBD requirements coming from SEL. In all possible directions where a parasitic SCRs can trigger a latch-up, an EGR or simply a biasing region cut the positive feedback loop.

3.5.3 Sensing

If DEMUX represents the most digital part of an SRAM, then sensing is the analog heart; there are several schemes that can differ a lot, but in principle, it is a sensing amplifier taking the two bit-lines belonging to the decoded cell in the array and evaluating, according to the bistable condition, at which level bit-line and negated bit-line are programmed.

Figure 3.44 shows a possible implementation of a sense amplifier. In this case, a standard differential stage with active load has been chosen with an output buffer giving strength to output signal of the sense amplifier. The input stage is represented by two p-channel transistors; usually, thanks to the higher mobility of carriers, the input stage is designed with n-channel but considering that under irradiation n-channel suffers TID degradation more than p-channel, a different approach has been adopted. The active load is built by a diode connecting n-channel transistor and another n-channel making the mirror.

The inputs of the differential pair are connected to the bit-lines of the cell, but considering that they can have connected a large number of cells, the distributed capacitance can be relevant; for this reason, bit-lines must be precharged and equalized to obtain the best condition to make the sense amplifier work properly. This is done by a pair of p-channel transistors, one precharging the bit-line and another one precharging the negated one. The precharge operation is usually done when the word-line and the bit-line of the cell to be read have been selected, and once completed the precharge phase,

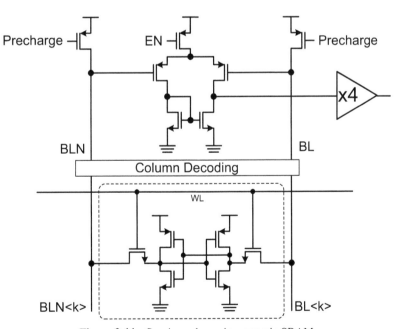

Figure 3.44   Sensing scheme in a generic SRAM.

the real sensing can be carried out to make the differential stage discriminate the bit stored in the cell.

The column decoding block of Figure 3.42 can be quite different and depend on how many bit-lines must be selected; for a low number of bit-lines, a single stage (represented simply n-channel transistors operating as switch) is enough, but for a high number of bit-lines, a pre-decoded terminal stage can be connected to another stage driving more elements. For a very large number of bit-lines (typical of large cuts), the stages can also be three, but in this case, the signal dynamic can become an issue in particular for rad-hard devices.

Figure 3.45 shows a possible physical implementation (layout) of a rad-hard sense amplifier. The first point to be underlined is that the implementation of EGR divides not only p-channel and n-channel transistor regions but also each single transistor. This is valid in particular for precharge transistors (on the left-hand side of the layout). The schematic shows one transistor only, but since the width is quite high when designing ELTs, it is always better to split the single transistor into three structures having the same W/L.

Being precharge transistors, the ones getting bit-lines in the best conditions for sensing, they are isolated from the sensing stage by EGR, and each transistor is also isolated by neighbour ones to guarantee stability also from SETs.

On the right-hand side of the layout, there is the differential pair with active load; the latter, built with n-channel transistors, has a double ring around to avoid any influence from p-channel transistors, while the p-channel differential pair is split into three transistors for left branch and three transistors for the right branch. As it is possible to see in the layout that the W/L

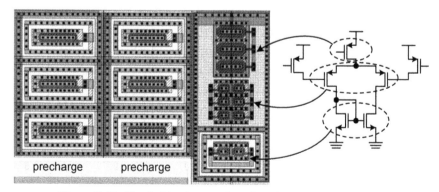

Figure 3.45   Physical implementation of a rad-hard sensing scheme.

of N and P transistors are equal, but there are three p-channels and only one n-channel: this to fit the mobility in NMOS, which is, in this technology, more or the less three times of PMOS.

The upper p-channel transistor, the one enabling the sensing scheme, is also split into three parts to obtain a better matching with the differential stage.

3.6 ECC Foundations: The Hamming Code

As mentioned in several parts of Chapter 2, RHBD cannot solve all problems related to silicon devices errors; hard ones can be mitigated at their minimum and a similar approach is valid for soft errors. ECC is the last bastion for rad-hard devices, and it is manly based on the concept of error detection and correction. It is important to underline that ECC is not an RHBD technique but a methodology to correct errors that can come from a charged particle upsetting a bit or any other event producing an error. ECC systems are valid, and indeed were born, to correct errors coming from communication lines, and sending a byte along a channel was an issue because of errors introduced by cables.

There are several ECC codes, from very simple ones to more complex and exhaustive, for very specific applications; in this chapter, the very basic Hamming Code is discussed just to make aware how it is not only a matter of implementing an ECC but how it deserves a rad-hard approach in the logic synthesis of the encoding and decoding schemes.

Table 3.1 shows a very canonical Hamming Code (12, 8), where parity bits are labelled as P (P0, P1, P2 and P3) and data bits (the byte) are labelled as D (from D0 to D7); the complete set of parity and data bit is called codeword, inside which parity bits are placed in positions corresponding to the power of 2: position 1 for P0, position 2 for P1, position 4 for P3 and position 8 for P3.

Calculating parity bits is quite simple and follows the concept of checking a bit (which can include both a data bit or a parity itself) and skipping another.

In Table 3.2, it is possible to see how P0 consider, among the codeword, a checking of the first bit (starting from itself) skipping the next one, checking

Table 3.1 Hamming Code (12, 8)

12	11	10	9	8	7	6	5	4	3	2	1
D7	D6	D5	D4	P3	D3	D2	D1	P2	D0	P1	P0

Table 3.2 Parity bit P0 calculation in Hamming Code (12, 8)

12	11	10	9	8	7	6	5	4	3	2	1
	D6		D4		D3		D1		D0		P0

the successive (D0), skipping the bit successive and so on until the end of the codeword. Thanks to bit positioning, as described in Table 3.1, this subset of the codeword (six bits in total) contains one parity bit and five data bits. Since data bits are known (these are the bits to be stored in a memory), it is a matter of calculating the parity bit P0, and the rule, according to Hamming Code, is to assign a "1" if the number of "1" contained in the five data bits is odd and a "0" if the number of "1" is even. This rule is quite simple and, on top of all, it represents an XOR function very easy to implement at circuit level with the corresponding logic port.

$$P0 = D0 \oplus D1 \oplus D3 \oplus D4 \oplus D6$$

Parity bit P1 follows the same rule of P0, this time not checking one and skipping one but checking two bits and skipping two bits, as shown in Table 3.3. In this case, the subset of the codeword is represented by six bits, where five are data and one parity. To calculate the parity bit P1, it is necessary again to count the number of "1" in data bits: if odd, P1 is "1", and if even, it is "0", according again to the XOR function:

$$P1 = D0 \oplus D2 \oplus D3 \oplus D5 \oplus D6$$

For parity bit P2, the rule is to check four bits and skip four bits (see Table 3.4) with the following XOR function:

$$P2 = D1 \oplus D2 \oplus D3 \oplus D7$$

For parity bit P3, being in the eighth position (see Table 3.5), the rule is to check eight bits and skip eight bits; since in the codeword the bits are only 12, only an XOR of the last four bits is considered:

$$P3 = D4 \oplus D5 \oplus D6 \oplus D7$$

The operation of generating parity bits starting from data bits is called encoding and, in the case of a Hamming Code (12, 8), it is composed by

Table 3.3 Parity bit P1 calculation in Hamming Code (12, 8)

12	11	10	9	8	7	6	5	4	3	2	1
	D6	D5			D3	D2			D0	P1	

Table 3.4 Parity bit P2 calculation in Hamming Code (12, 8)

12	11	10	9	8	7	6	5	4	3	2	1
D7					D3	D2	D1	P2			

Table 3.5 Parity bit P3 calculation in Hamming Code (12, 8)

12	11	10	9	8	7	6	5	4	3	2	1
D7	D6	D5	D4	P3							

four logic net using XOR ports. According to Chapter 2, RHBD digital cells can be used to obtain an encoding rad-hard engine, and using simple XOR ports with two inputs (XOR with 10T architecture to maximize the number of p-channel transistors connected to power supply and n-channel transistors connected to ground), it is possible to realize a radiation hardened net like the one shown in Figure 3.46.

Once both data and parity bits are encoded and stored, the codeword is ready to be read, and in this case, a reverse operation must be done to check if a bit has been upset. Indeed, a charged particle can hit either a bit data or a parity bit but thanks to Hamming Code whenever the error is in the first or in the latter, it can be detected and corrected.

The reverse operation is done again by making leverage on XORs, and this time, all codeword bits are used to extract four detected parity bits $DP<0:3>$:

$$DP0 = D0 \oplus D1 \oplus D3 \oplus D4 \oplus D6 \oplus P0$$

$$DP1 = D0 \oplus D2 \oplus D3 \oplus D5 \oplus D6 \oplus P1$$

$$DP2 = D1 \oplus D2 \oplus D3 \oplus D7 \oplus P2$$

$$DP3 = D4 \oplus D5 \oplus D6 \oplus D7 \oplus P3$$

Since embedded SRAMs usually have different pins for input and output bits, the above described XOR operations may be translated in the scheme of Figure 3.47, where DOUT and POUT are retrieved by the codeword.

Detected parity bits give not only the detection of the error but also its position according to Table 3.1, where data and parity bits are placed.

Table 3.6 shows an example of codeword coming from 01010101 (checkerboard) for data bits; parity bits are 0010.

Table 3.7 shows the change of a bit (D1) coming from a charged particle; it can happen from a transient in a terminal stage or from an upset occurred

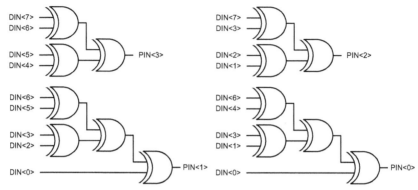

Figure 3.46   Logic network implementing Hamming (12, 8) encoding for parity bits.

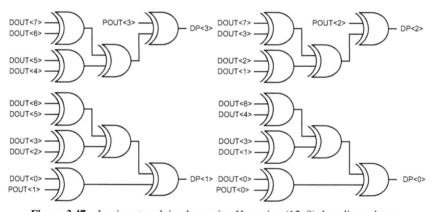

Figure 3.47   Logic network implementing Hamming (12, 8) decoding scheme.

at cell level. Assuming that the bit has been changed from "1" to "0" when codeword is read, the reverse algorithm is applied as shown in Table 3.8.

The detected codeword shows two errors in parity bits check: the first parity bit (P0) and the third (P2). Since their positions, according to Table 3.8, are the first and the fourth, their sum tells that in the fifth position, corresponding to bit D1, there is the wrong bit.

Table 3.6   Codeword example

12	11	10	9	8	7	6	5	4	3	2	1
D7	D6	D5	D4	P3	D3	D2	D1	P2	D0	P1	P0
1	0	1	0	0	1	0	1	1	0	0	0

Table 3.7 Codeword example, where one data bit has been upset

12	11	10	9	8	7	6	5	4	3	2	1
D7	D6	D5	D4	P3	D3	D2	**D1**	P2	D0	P1	P0
1	0	1	0	**0**	1	0	**0**	1	0	**0**	**0**
							SEU				

Table 3.8 Parity bits detection and comparison

12	11	10	9	8	7	6	5	4	3	2	1
D7	D6	D5	D4	P3	D3	D2	D1	P2	D0	P1	P0
1	0	1	0	0	1	0	0	1	0	0	0
				0				0		0	1
				OK				KO		OK	KO

The implementation of data recovery starting from the codeword is, at this point, quite trivial and it is in principle the reverse operation done for generating the codeword itself. Since parity bits are four, there are 15 possible conditions, where the first one (Q<0>) is the no-error condition and it is easily implementable with a 4–16 decoder by using simple four inputs AND ports. In the example shown, eight data bits have been used, but four parity bits can serve up to 11 data bits, and for this reason, in the network shown in Figure 3.48, the last three positions (Q<13>–Q<15>) in the decoder 4–16 are discarded and only eight bits are reconstructed starting from the data out of the SRAM.

Figure 3.49 shows the floor-plan of a generic SPSRAM implementing the Hamming ECC described above. Data bits are eight (from D0 to D7) and parity bits are four (from P0 to P3). Each bit has its own decoding scheme (DEMUXs, terminal drivers and read and write systems). In the

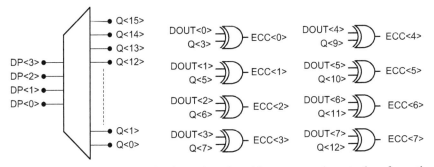

Figure 3.48 Logic network implementing data bit reconstruction starting from the codeword.

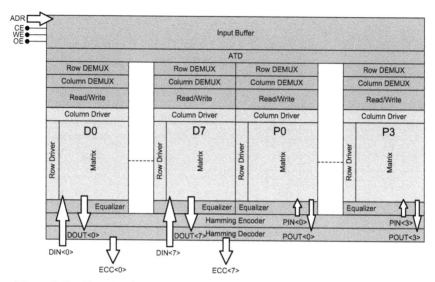

Figure 3.49 Complete floor-plan of a rad-hard SRAM using a Hamming (12, 8) ECC.

bottom of the scheme are sketched Hamming Encoders and Decoder, where it is possible to see how external data to be written (DIN<0:7>) access directly the SPSRAM writing data without any change from ECC systems; concerning parity bits PIN<0:3> are generated by the encoder and stored in four bits bank, while detected parity bits POUT<0:3> are provided directly to the Hamming Decoder together with data out bits DOUT<0:7>. The data out come from the decoder (ECC<0:7>) and represent the real corrected data by the ECC engine. Optionally, it is also possible either to get out parity bits or directly the decoder 4–16 outputs to make aware externally if the ECC engine has corrected the byte. This can be useful to understand how much the ECC engine is working and if it necessary to make additional operations like scrubbing.

An SPSRAM built with the floor-plan like the one described in Figure 3.49 represents a real rad-hard device; the architecture making leverage on independent peripheries guarantees only SEUs and the Hamming (12, 8) ECC guarantees the correction of single errors.

But this can be not enough. A bit may be hit also when the memory is in stand-by and even if it can be corrected at the first read, it can happen that a cumulation of errors may occur; this progressive SEUs cumulation can involve more bits in the same byte and in this case the ECC algorithm cannot make any correction. This situation is quite rare in small memories, but it strongly depends on how much frequently the memory is written; if there are frequent write and read operations on the whole array, the probability to have a cumulation of errors on the same byte is low and ECC can make its work in an efficient way, but if memory banks are written and stand-by last long enough, error cumulation can become an issue.

The only way to solve this problem is to rewrite the entire memory or rewrite the byte if the device is aware that the last read has been fine only because ECC engine has corrected the byte. Of course, rewriting data each time the ECC engine makes a correction means stopping the read operation, enter in write mode and rewrite both data and parity bits and come back in read mode. This can be a problem in terms of access time, and data flow to and from a core processor can become somehow painful.

Many ECC algorithms exist to solve this problem; an SECDED (Single Error Correction, Double Error Detection) algorithm, for example, can help and if used in a memory device that is known to be safe enough in terms of MBUs when the engine starts to intercept double errors (DEDs), it may mean that a cumulation of errors occurs and it is time for a scrubbing operation. When and how to make working scrubbers is something very strictly related to ECC and is a combination of data safety and performance of the overall devices, but this is not RHBD anymore: it is error handling!

References

[1] A. Stabile, V. Liberali, C. Calligaro, 'Design of a Rad-hard library of digital cells for space applications', IEEE International conference on electronics, circuits and systems 2008 (ICECS), pp. 149–152, September 2008.

[2] M. Benigni, V. Liberali, A. Stabile, C. Calligaro, 'Design of Rad-Hard SRAM Cells: a comparative study', Proceedings 27th International Conference on Microelectronics (MIEL) 2010 (Best Paper), Serbia.

[3] C. Calligaro, A. Arbat Casas, Y. Roizin, D. Nahmad, "A 15 Mrad (Si) 512 Kbit Rad-Hard SRAM in a standard 0.18um CMOS technology", RADECS 2012.

[4] C. Calligaro, V. Liberali, A. Stabile "A Radiation Hardened 512 Kbit SRAM in 180 nm CMOS Technology", 16th IEEE International Conference on Electronics, Circuits, and Systems (ICECS), Hammamet (Tunisia), Dec. 2009, pp. 655–658.

[5] A. Arbat Casas, C. Calligaro, Y. Roizin, D. Nahmad, 'Radiation hardened 2 Mbit SRAM in 180 nm CMOS technology', October 2012, 2012

IEEE First AESS European Conference on Satellite Telecommunications (ESTEL).

[6] C. Calligaro, V. Liberali, A. Stabile, M. Bagatin, S. Gerardin, A. Paccagnella, "A multi-megarad, radiation hardened by design 512 kbit SRAM in CMOS technology", 22nd International Conference on Microelectronics (ICM 2010).

[7] A. K. Sharma, "Advanced Semiconductor Memories", IEEE Press – Wiley Inter-Science. ISBN 978-0-471-20813-6.

4

One-Time Programmable Memories for Harsh Environments

Umberto Gatti

RedCat Devices, Italy

"The people who can destroy a thing, they control it."

Frank Herbert, Dune

One-Time Programmable (OTP) Memories belongs to the big family of the Non-Volatile Memories (NVM). Even if their early development dates back to the early 1980s, they are gaining in popularity due to various added benefits including cost reduction, improved performance, enabling secure storage and easy configurability. This applies also to space market, where OTPs are widely used on board the satellites. In this chapter, we will review the basic concept of OTP Memories and the reasons why it is worth focusing on those based on anti-fuse in space applications. Moreover, we will discuss which provisions shall be taken during their design to mitigate the radiation effects.

4.1 Introduction

One-Time Programmable (OTP) Memories belongs to the big family of the Non-Volatile Memories (NVM). Even if most of them are not suitable to be directly used in a harsh environment, they can be the starting point for the development of a rad-hard counterpart. In this chapter, we will focus on the OTP since they represent the most economic device that can be utilized on-board the satellites and the space market is currently eager of them.

4.1.1 NVM Technology Overview

In the arena of the embedded NVMs for general applications, there are several technologies commercially available. The NVMs can be categorized in different ways. One is on the basis of the principle used to store the logic information "0" and "1" in the elementary cell of the array (bit-cell) [1]. A bit-cell can include either only the storage element or the storage element plus the selector. A brief list (not exhaustive) is the following [2]:

- Embedded flash based on either floating-gate or split-gate architectures;
- Floating gate, which uses hot carrier injection for programming;
- Read-only Memories (ROM), where the logic values "0" or "1" can be determined through a via mask;
- Electrical fuse (called also eFuse), where a polysilicon or a metal line is blown to create a change in its resistance;
- Anti-fuse (AF), where the memory is programmed by means of a hard oxide breakdown of a gate or a poly layer to create a change in its resistance.

Further NVM, developed with new materials, are in the early stage of development but are beyond the scope of this chapter.

Flash memory working principle is the modification of the threshold voltage of an MOS transistor by trapping charge in a storage layer added to a traditional transistor. Embedded flash memories are easily reprogrammable, so that they are used for the storage of codes that change often, for example, in microcontrollers (MCU) or in a system-on-chip (SoC) to provide flexibility to end applications. This flexibility is paid in terms of a higher cost, because embedding flash memories adds extra mask layers into logic circuit fabrication processes (up to 10 additional layers).

Floating gate memories are specific memories, usually integrated at front-end level, which can be used as OTP. Thanks to their floating gate, which can store charges, they can be written and erased many times. These are old technologies known since the 1970s as EPROM (Erasable PROM) or EEPROM (Electrically Erasable PROM), depending on the programming and erasing mechanism (hot carrier injection for EPROM and Fowler–Nordheim tunnelling for EEPROM).

ROMs are used for the storage of a fixed code, such as in microprocessors or in computer to store the BIOS. It is the least expensive of the embedded NVM technologies, as well as the least flexible. The programming of the ROM is done only in fabrication by means of the mask which determines the connections between two metal layers. Once the IC is fabricated, the content cannot be changed without a mask re-spinning.

On the other hand, the other technologies, such as anti-fuse and eFuse, can be programmed once either in-fab or in-field after manufacturing. In the AF, the memory cell is based on a thin oxide sandwiched between two electrodes and data writing is made breaking the thin oxide by applying to it a voltage pulse. In the eFuse, instead, the cell is based on a thin poly (or metal) line that is broken by Joule effect during programming by a current pulse.

Table 4.1 shows a comparison between available NVM with the different technologies based on cell structure, process compatibility, bit-cell areas, security, scalability, stand-by and active current, access time and voltage/temperature tolerance [1–3].

It can be noticed that OTP based on anti-fuse and ROM show similar performance, but the former offers one more degree of freedom in programmability and increased security.

Table 4.1   Embedded NVMs comparison

	Embedded Flash	ROM	Floating Gate (OTP)	Fuse (OTP)	Anti-fuse (OTP)
Cell architecture	Floating gate 1T/2T	Masked 1T	Floating gate 1T/2T	Metal or poly fuse	Anti-fuse 3T, 2T, 1.5T, 1T-1C
Process	Flash	CMOS	EPROM/ EEPROM	CMOS	CMOS
Standard CMOS compatibility	No	Yes	Yes	Yes	Yes
Bit-cell area (normalized)	3	<1	10	300	1
Scalability	☺	☺	☹	☺	☺
Additional actions	Up to 10 masks	–	UV Erase	–	–
Stand-by & active consumption	☺	☺	☹	☹	☺
Access time	☺	☺	☺	☹	☺
Security	☺	☹	☺	☹	☺
Programmability	In-field	In-fab (mask)	In-fab/ In-field	In-fab/ In-field	In-fab/ In-field
Temp & Voltage Tolerance	☹	☺	☹	☺	☺
Radiation Resiliency	☹	☺	☹	☺	☺

As we will see later, one key factor in the choice of the most suitable solution in a memory for space is its manufacturability in a standard CMOS process, to whom it is possible to apply RHBD techniques to make it resilient to radiations. This in turn allows easy portability, scalability, maintenance and embedding capability.

4.1.2 OTP Application in Harsh Environments

In all electronic systems, including the equipment for space applications, semiconductor memories play two fundamental functions that require different memory technologies. The first one is the temporary storage of data used to perform the different functions of the system, where it is not necessary that data are retained when the power supply of the system is switched off. The other important function covered by memories is the permanent storage of code, tuning parameters and other data that do not need to be changed or that are updated very few times during the operative life of the system. The nature of the data stored requires that they are not lost when the power of the system is removed, and to reach this result, an NVM is used. The key factor in the choice of a suitable NVM for space applications is its resiliency to radiation effects. In respect to this, flash memory and EPROM/EEPROM show a radiation softness that is intrinsic to their technology and not so easy to overcome: the storage mechanism of the bit-cell is based on charge storage in a floating gate. Under the impact of high-energy radiation, enough energy can be transferred to some electrons to jump over the oxide barrier and defect can be created in the oxide generating traps or charge can be trapped in the oxides all around the floating gate bending the conduction band and increasing the charge tunnelling probability. Flash memories can suffer also from Single-Event Functional Interrupt (SEFI), which is a soft error that causes the component to reset, lock-up or otherwise malfunction. SEFI typically occurs in complex devices with built-in state/control sections like in modern memories (NOR- and NAND-Flash, etc.).

On the contrary, anti-fuses, eFuses and ROM are intrinsically not sensitive to radiations. The question is if preferring OTP with respect to ROM makes sense. ROM cells are small and, once programmed, are irreversible and hence reliable. However, they have two major shortcomings: low security and lack of in-field programmability:

- Security: because a masked ROM is hardwired during chip fabrication, it is relatively easy to reverse engineer to see the programming of an ROM block. The states of the individual bit-cells, and hence the code stored in ROM's, can be extracted using an SEM (scanning electron microscope).

- In-Field programmability: the other major problem with ROM is that it is programmed during wafer processing and cannot be modified afterwards. If used for code storage, this means that code must be frozen when silicon is ready, which eliminates the possibility of modifying this code after silicon verification in the target system.

So, the answer to the question is positive: an OTP based on anti-fuse or eFuse is an excellent choice for electronic equipment used in harsh environments.

OTP have been generally employed in a number of applications due to their low-cost manufacturing and full compatibility with CMOS. The main applications are:

- Code storage for security or cryptography;
- Embedded-RAM repair (the increase in manufacturing variability with advanced technologies processes) and
- Analogue block trimming: laser zapping is no more suitable for advanced technologies, so fuses/anti-fuses represent an efficient alternative.

In addition, especially in space application, OTPs are widely used for the boot software of the operating systems of the satellites, for storage of stellar maps, to store the bit-code of any rad-hard FPGA and for all those applications that require fixed data written to the ground and simply read in flight.

On-Board Computer

The spacecraft control computer is a key element for controlling the spacecraft and maintaining its safety. The control computer, whose internal architecture may depend on the application and its requirements, includes autonomous failure management functionalities to ensure that the spacecraft can autonomously recover from major anomalies and enter a safe state without interaction from the ground operators.

The term "On-Board Computer" (OBC) indicates any unit flying on board a satellite which provides processing capability. However, the OBC more commonly refers to the computer of the satellite's avionic sub-system, i.e. the unit where the on-board software runs. In turn, the "on-board software", despite the rather generic label, is known as the software implementing the satellite's vital functions such as attitude and orbit control, tele-commands execution, housekeeping telemetry gathering and formatting, on board time synchronization and distribution, failure detection and isolation and recovery.

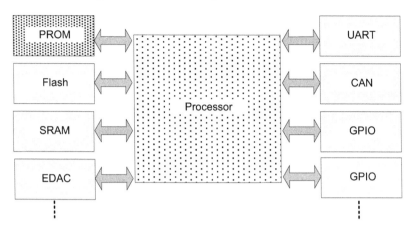

Figure 4.1   Simplified system block diagram of a typical OBC.

Based on the above, the very essence of an OBC is the microprocessor board, consisting of microprocessor, NVM, volatile memories and the interfaces chips that connect the microprocessor to different peripherals [4]. Figure 4.1 shows the simplified system block-diagram of an OBC. It is a general-purpose computer for space applications. It features a program memory which is protected by error detection and correction (EDAC) and boot system storage on PROM, which is an OTP.

Stellar Maps

Stellar maps, stored in OTP, are used in conjunction with star tracker device that measures the position of star using photocells or a camera. As the positions of many stars have been measured by astronomers to a high degree of accuracy, a star tracker on a satellite or spacecraft may be used to determine the orientation (or attitude) of the spacecraft with respect to the stars. In order to do this, the star tracker must obtain an image of the stars, measure their apparent position in the reference frame of the spacecraft and identify the stars so their position can be compared with their known absolute position from a star catalogue. A star tracker may include a processor to identify stars by comparing the pattern of observed stars with the known pattern of stars in the sky stored in memory [5].

OTP for FPGA Booting

During a satellite operating life, the tasks to be performed can evolve and the protocols may change. Therefore, the flexibility offered by reprogrammable FPGAs is particularly appreciated to dynamically reconfigure the satellite

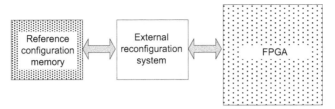

Figure 4.2 General architecture of the recovery architecture for an FPGA [6].

hardware during the mission. However, such high levels of radiation affect the reliability of the electronics and may produce undesired alterations of the data being processed, of the state machines status and of the stored data and even damage the functionality of full digital blocks. Regardless of the specific solution implemented to prevent a particular fault type, all satellites adopt a basic safety procedure based on the periodic reboot and reload of the original FPGA configuration called scrubbing. The FPGA bit-stream to be reloaded can be stored in different memory types, one of these is anti-fuse.

The general architecture of a configuration recovery system is illustrated in Figure 4.2. An external configuration memory (SRAM, flash or anti-fuse) stores the reference configuration information and is read by the scrubber block that supervises the refresh procedure and controls the loading of the bit-stream into the FPGA. The scrubber block can periodically refresh the FPGA.

4.2 OTP Memories for Standard CMOS Technologies

As mentioned in Section 4.1.1, general-purpose OTP memories can be based on fuses, anti-fuses or floating gate (EPROM/EEPROM). However, the latter tends to reduce their diffusion since data retention on floating gate becomes a reliability issue for reduced oxide thicknesses unavoidable in advanced technologies. Moreover, they are prone to radiation effects, which make them unsuitable for space application. For this reason, in the following we will focus on fuses and anti-fuses only.

The challenges for new OTP memories are high density, high reliability, high capacity, lower programming voltage, lower programming current, speed at least as good as standard flash memories and compatibility with standard CMOS process for SoC development, i.e. no additional mask for low-cost manufacturing.

4.2.1 Principle of Operation

Fuses and anti-fuses are based on the alteration of the physical property of a material [7]:

- A fuse is implemented with an electrical conductor, which is blown during programming.
- An anti-fuse is implemented with a material alterable to a conductive state by programming.

So, whereas a fuse starts with a low resistance and is designed to permanently break an electrically conductive path (typically when the current through the path exceeds a specified limit), an anti-fuse starts with a high resistance and is designed to permanently create an electrically conductive path (typically when the voltage across the anti-fuse exceeds a certain level). Figure 4.3 shows a simplified view of the electrical state changes from resistor to capacitor by programming and vice versa [8].

The working principle of an electrical fuse is the fusion of a conducting layer, which can be metal or salicided polysilicon. Polysilicon fuses have been widely used for several generations of CMOS technologies starting in the late 1990s over a decade. A polysilicon line heats up to fusion by self-heating mechanism due to electrical conduction. The conduction current controls this phenomenon. Localization of the fusion point requires a specific design of the polysilicon link, as shown in Figure 4.4 [7,9] in light gray. Fusion of this link increases its resistance and eventually opens the connection. This resistance variation conditions the sense current that is compared to a reference current. Difference between them is then evaluated as logic one or zero. The fuse element is composed of a thin layer of gate-material silicide. To program the fuse, a high current was passed through the fuse leading to the agglomeration of the lower-resistance silicide.

Poly fuses have usually short programming times in the microsecond range and low programming voltage, but unfortunately, they require large

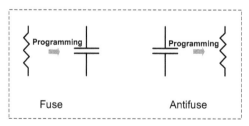

Figure 4.3   Concepts of electrical state changes by programming on anti-fuse and fuse.

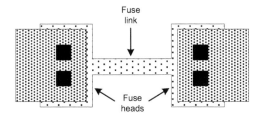

Figure 4.4   Generic layout of a polysilicon fuse.

programming currents in 10mA range [10]. Thus, they exhibit poor density at bit-cell level, thus preventing the implementation of large memory array. Moreover, the stored code can be easily identified with reverse engineering.

On the other hand, metal fuse was initially implemented with a Cu trace or a Cu-via interconnection. The non-programmed metal trace is a short circuit. Fusion of Cu necessitates a large programming current and large programming time. The bit-cell is quite large and limit array sizes. The interest on metal fuse has recently re-vamped when the industry transitioned to high-k metal gate CMOS technologies [11, 12]. Since the bit-cell includes also a selection MOS, it is also called 1T-1R by the authors. This device is easily scalable to future logic technologies. Moreover, metal-fuse technology allows the element to be stacked above the access transistor (3D stacking, see Figure 4.5).

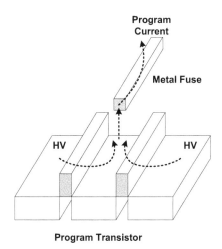

Figure 4.5   Fuse and selector MOS stacking enabled by metal-gate technology [11].

Anti-fuse is an electronic device that changes state from non-conducting to conducting or from higher resistance to lower resistance in response to an electrical stress (programming voltage or current) [13].

First anti-fuses date back to the late 1960s, with inter-metal capacitor blown at the row/column cross point. At present, the most typical and mature anti-fuses combine a selection MOS with an oxide layer, which can be ONO (Oxide–Nitride–Oxide) type [14], α-Si (Amorphous Silicon) type [15] and gate-oxide type. Compared with ONO and α-Si anti-fuses, gate oxide anti-fuse has its advantage: it is entirely compatible with the standard CMOS core technology and it scales naturally with technology. Gate-oxide AF performance corresponds to the ONO type's and α-Si type's and it can cut the cost greatly. The design procedure also becomes more convenient and rad-hard techniques can be applied easily. Indeed, MOS capacitor is an excellent anti-fuse and is programmed by gate oxide rupture, causing a resistive change. Typical characteristics are:

- gate leakage is in order of pA or nA;
- oxide breaks down at 5–8 V and
- gate current increases >10,000x

Figure 4.6 represents a typical gate-oxide capacitor. The exact mechanism is still not well understood; however, an explanation can be the following.

From the physical point of view, when a programming voltage, VPP, is applied, the oxide breakdown occurs in two stages [7]: a phase of "tunnelling" or (Fowler–Nordheim), which creates an accumulation of charges and a conductive path through the oxide (the electrical fields present can be of the order of a few tens of MV/cm), and a second step of local heating as a result of the passage of the current and the increase in temperature (thermal runaway), which creates a "dissolution" (melting) in the oxide (hard breakdown).

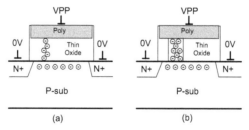

Figure 4.6 Typical gate-oxide breakdown in AF during programming, with two phases [7, 13]: (a) conductive path creation and (b) thermal runaway (melting).

The AF suffers a real structural change that allows that a significant current can flow between gate and source/drain.

While the gate-oxide employed in the AF device is usually thin, the selection transistor, which is usually in series with the anti-fuse device, is realized with a thicker gate oxide thus preventing breakage during programming phase. Currently, thanks to the continuous scaling in the thickness of gate oxide, which is much faster than the supply voltage scaling, it is possible to produce AF based on sufficiently thin oxides, which do not need very high programming voltage.

4.2.2 CMOS OTP Based on Anti-Fuse

As seen, the OTP are among the simplest and economic non-volatile memories. Those based on anti-fuse represent a low-cost solution for embedded applications. Their advantage lies in the easy programmability with low currents both during manufacture (in-fab) and during use (in-field), without the need for additional masks or steps of dedicated manufacturing. This characteristic makes them easily portable between different manufacturing sites and between different processing nodes. Figure 4.7 reports the taxonomy of the OTP memories, where those based on AF will be discussed in detail.

One possible cost-effective realization of a single memory cell (bit-cell) anti-fuse is based on gate oxide and uses the same oxide breakdown, which is

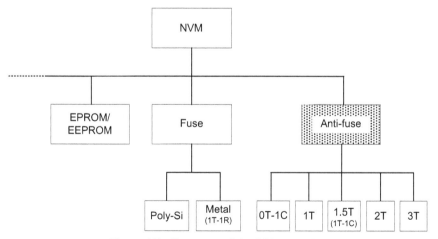

Figure 4.7 Taxonomy of the OTP memories.

normally an insulator, as a means for storing a bit. In general, the gate oxide is chosen because it is available in standard CMOS processes, can scale with technology, while the oxide quality is under control (thickness, purity, etc.). Moreover, being thin can be broken quite easily. The programming procedure depends heavily on the oxide thickness. After the oxide breakdown AF realizes a permanent conductive path between the gate and the source/drain and behaves like a resistance. The resistance value depends on the programming voltage used for the break-up (and therefore also from the current) and the time for which this voltage has been applied.

There are several topologies for the implementation of a memory cell based on an AF gate oxide. The AF and one or more transistors for selection, connected in series, are the two basic elements of the bit-cell. There are cells made of 0T-1C (no selection transistor) [16], 1T (one transistor), 1.5T (called also 1T-1C), 2T (two transistors) or 3T (three transistors). The 0T-1C needs four additional masks with respect to standard CMOS process, so we do not discuss it.

Since the early 1980s, there have been technological solutions for the realization of OTP-based AF. For example, already in [17], the AF principle based on the gate-oxide breakdown was presented. However, since in old technologies the oxide breakdown voltage was much higher than the supply, it was mandatory to introduce a modification to the standard manufacturing process (called "NMOS drift" or "1Tdrift- 1C" in [17]) to make the selection MOS more robust. Currently, thanks to the continuous scaling in the gate oxide thickness, which is much faster than the scaling of the supply voltage, it is possible to realize AF based on a sufficiently thin oxide, which does not require very high programming voltage.

The benefits of anti-fuse solutions based on standard CMOS technology can be listed as:

- Manufacturability for mass production and yield
  - No additional process control required
  - Standard MOS transistor
  - No additional wafer processing operations
    - No wafer bake
    - No UV erase
- Same reliability as foundry process
  - Standard CMOS baseline logic process technology
  - Exclusive use of standard V_{th} MOS devices

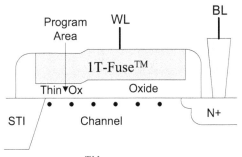

Figure 4.8 Patented 1T-Fuse^{TM} Split-Channel Bit Cell (Sidense, [18, 19]).

The only additional requirement is dual oxide thickness (thin and thick) process for AF device and selector MOS, respectively. However, all recent CMOS technologies have two kinds of oxide thicknesses for internal logic circuits and I/O circuits, because of higher interface voltage than core voltage.

The 1T topology was proposed by Sidense [18], and is also called Split Channel Device (Figure 4.8). The 1T-Fuse^{TM} bit cell is a two-terminal, split-channel device that looks like an MOS capacitor in the un-programmed state and a diode-connected MOS transistor in the programmed state. All programming occurs in the transistor's channel region. It consists of a thin (core) gate and a thick gate. The thin gate is the program gate, while the thick gate is the select gate.

When a normal supply voltage is applied to the word-line (WL), no current is sensed along the bit-cell. The equivalent circuit is a capacitor. Since there is no current that flows along the bit-line (BL), the bit-cell is "0" by default. When a large programming voltage is applied along the gate of the program transistor, a hard gate oxide breakdown occurs, and a resistive path is created. Due to the thickness gradient, hard oxide breakdown occurs at the weakest link, the junction of the thick gate and thin gate. It can be manufactured in standard-logic CMOS processes and does not require any additional mask layers or process steps.

The 2T bit-cell in [3, 20] is constituted by two nMOS, one used as AF (with a thin-oxide), in which drain and source are shorted together and connected to the source of the selection MOS MS (using thick-oxide), while the gate is connected to the high voltage. The cell has three terminals and an area equal to that of two complete transistors (Figure 4.9).

Before programming, the AF can be modeled as a capacitor. It changes to a resistor after gate oxide breakdown, representing two memory states.

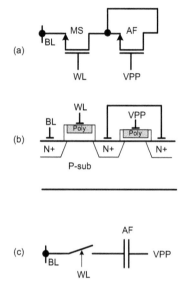

Figure 4.9   Anti-fuse 2T cell (a) Scheme, (b) Cross-section, (c) Equivalent circuit.

A programmed AF conducts current on the bit line, which is detected by the sense amplifier. A pMOS counterpart can be found in [21]. During the programming phase of a cell, the word-line is on (polarized, for example, around +3.3 V for a CMOS 180 nm process), the selected bit-line is fixed to ground, while VPP (usually from 5 V to 8 V for the same technology) is driven by a train of high voltage pulses and the duration of which helps to determine the value of the programmed resistance. When the VPP programming voltage is applied and the oxide breaks down, a high voltage is propagated on the drain of the selection transistor MS causing a stress, with an impact on reliability.

The solution 3T [22, 23] uses a cascode type solution for the selection of the bit-cell, so that the high voltage is divided between two selection transistors (Figure 4.10). During programming, the voltage on the drain of MOS MS (with thick oxide) is limited to VB-V_{th}. Of course, the area is greater than the 2T cell and the cell has four terminals. A pMOS version of the same anti-fuse can be found in [24].

A further and more recently presented solution makes use of a transistor and a half and is defined 1.5T [25, 26]. The 1.5T cell is composed of an access transistor coupled in series with an AF device, as shown in (Figure 4.11). The selection transistor addresses the cell to be programmed

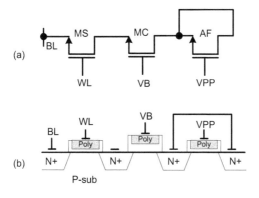

Figure 4.10 Anti-fuse 3T cell (a) Scheme, (b) Cross-section.

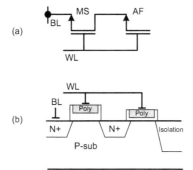

Figure 4.11 Anti-fuse 1.5T cell with shared gates (a) Scheme, (b) Cross-section.

and read or not. The AF device (an nMOS capacitor) uses the breakdown of gate oxide as a storage node by acting as an equivalent resistance between gate and source. The effective program voltage strongly depends on the thickness of the gate oxide. By sharing the source of the AF device and the drain of the access transistor and shorting their gates, the size of the cell is small. The gate of MS and AF can also be kept separated as in Figure 4.12 for higher flexibility. In both cases, a high-voltage drives the gate of AF.

The cell allows a reversible area reduction of about 50% compared to the 3T solution and a considerable simplification in routing within the memory array. Two different gate oxides, thin and thick are used in the 1.5T bit-cell. Thin oxide is used in the AF to facilitate its breakage, and the thick one in the selection MOS is to make it more reliable during the high-voltage programming phase. The high voltage is applied directly to WL in Figure 4.11

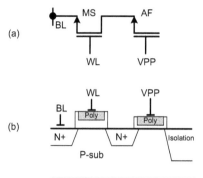

Figure 4.12   Anti-fuse 1.5T cell with split gates (a) Scheme, (b) Cross-section.

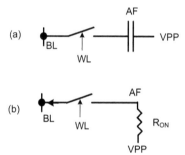

Figure 4.13   Anti-fuse 1.5T cell equivalent circuit (a) Before programming, (b) after programming.

or through final stages in Figure 4.12. Accordingly, also the final stages of cell decoding circuitry shall be made with MOS with thick oxide. Typically, the thin oxide thickness can be around 3 nm, whereas the thick oxide is often around 6–7 nm for 180 nm CMOS process.

The equivalent circuit in the two phases (before programming and after programming) is shown in Figure 4.13.

4.2.3 Characteristics and Limits

An anti-fuse bit-cell is specified by several suitable characteristics. The programmed resistance value, R_{ON}, is one of the critical characteristics for the AF, since in a complete OTP, the reading characteristics, such as the access time and sensing margin, are mainly dependent on that value. This value can be varied with the breakdown position (i.e. where the oxide breakdown spot is located). However, the most dominant factor on the value of R_{ON}

is the supplied current when the AF breaks down. In turn, the amount of supplied current during the AF breakdown depends on the size of selection MOS [27]. Related to this, there are two other important parameters: R_{ON} value distribution and AF read window (or sense margin).

Indeed, gate-oxide anti-fuse memory cell also has a very typical weakness, that is, the widespread nature of the programmed cell. The equivalent resistance of memory cell distributes from several hundred ohms to several hundred kilo ohms. This defect increases the difficulties in designing the sense amplifier. The read speed of OTP is also facing a big challenge. If the equivalent resistance is too large, the circuit may even fail. In Figure 4.14, the distribution of both cells with high resistance R_{OFF} (corresponding to un-programmed AF, whose ideal equivalent circuit is a capacitor) and cells with low resistance R_{ON} (corresponding to programmed AF) is reported. Sometimes in the literature, instead of the R_{ON} value distribution, the AF current, I_{AF}, distribution is reported. Both distributions (fitting approximately a Gaussian) have a spread due to geometrical parameters, process, temperature and programming voltage/time variations and it is evident that to discriminate bit "1" and bit "0", these distributions must be separated: the space between the two distributions is called read window. Thus, it is evident that the read margin shall be improved in some way for safe operation.

It is well known that read operation of an AF requires the identification of the R_{ON} value by means of the current I_{AF} flowing across R_{ON}, or transforming it in a voltage, V_{AF}. A comparison between I_{AF} or V_{AF} and a reference value (V_{ref} or I_{ref}) implies that the threshold shall be put in the middle of the read window. It comes out that the available window for each data is actually 1/2 of the total read window, but this window is further reduced considering that a guard-band Δ must be put at each side of the

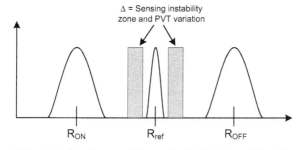

Figure 4.14 Anti-fuse R_{ON} and R_{OFF} distribution for a 1bit-cell memory.

reference value to take care of the sensing amplifier instability zone due to circuit non-ideality and process–temperature–voltage (PVT) variations.

A first method to improve the read window is by adopting a differential approach that makes the reading more robust (see also Section 4.3.3). The concept is to use two cells for a single bit and comparing each other after programming them with a complementary data, thus eliminating the reference cell with middle point threshold.

There are two main drawbacks related with this solution: the first one is a heavy area consumption; a physical 512kbit physical array is necessary to obtain a logical 256kbit device. This overhead should be unacceptable for a large memory size, but it is acceptable for medium-low memory size. The second problem is that the writing operation may require a longer time than a standard approach. Also in this case, considering the reduced memory size and the fact that the operation is typically done one time at customer location this extra time is definitively acceptable.

A second approach is to maximize the difference in value between the resistance presented by the programmed cell, R_{ON}, and that one which is not programmed, R_{OFF}. With standard CMOS technology, the expected median value of the Gaussian distribution, R_{OFF}/R_{ON} ratio, should be >1000. However, this ratio is closely associated with the programming conditions. Soft breakdown (an incomplete melting of the oxide) shall be absolutely avoided, since it can produce an R_{ON} value of the order of 1–10 Mohm [28], easily mistakable with that of an un-programmed bit-cell. But also a complete hard breakdown can produce R_{ON}, with median value of some kohm and a distribution with 3σ of 2 orders of magnitude (up to some hundred kohm).

The programming process has a direct impact on the breakdown evolution. It depends first on the thickness of the oxide. As said, the resistance value R_{ON} depends on the programming voltage used for breaking (and therefore also on the current) and the time that this voltage has been applied. For example, in [29], it was observed that a 20% reduction of programming current degrades the anti-fuse current by approximately a half order of magnitude. Doubling the programming current increases the anti-fuse current by approximately 1 order of magnitude. To increase the programming current, i.e. transistor size of the write-port, it is necessary to enlarge the memory cell area size.

Programming time has also an impact on energy consumption and the testing time. Consequently, the programming time should be minimized. The programming voltage value is another game-player since it determines the final value of the programmed resistance. It brings also reliability concerns.

Finally, the memory density is dictated by the bit-cell size that comprises the area of the anti-fuse device and the channel dimensions of the access transistor. Thus, deep understanding of gate-oxide breakdown mechanism and post breakdown device characterization is essential to obtain high yielding OTP devices. Small lots of cells shall be foreseen in order to identify the optimized size of selector MOS and characterize as much as possible the AF bit-cell behaviour before production [27].

4.3 Rad-hard CMOS OTP

4.3.1 Critical Features

As usually happens when using standard CMOS technologies in space and equivalent harsh environments, OTP memories are exposed to radiation effects. Concerning Total Ionizing Dose (TID), scaling of CMOS technologies brought along an increased level of radiation hardness, thanks to gate oxide thinning: in thin oxides (few nm), radiation-induced positive charge reduces. Low voltage, which means lower electric field, helps to further reduce TID effects. However, lateral isolation oxides were (and are) still thick and radiation-soft. Thus, intra- and inter-device leakage are still there, above all, in selector MOS, which use only thick oxide. As for Single Event Effects (SEE), the reduced critical charge and higher density make the devices more sensitive to high-energy ion hits.

Effect of radiations in a circuit shall be analyzed at cell level, at array level and on the global memory, including periphery (buffers, demuxes, timing circuit, etc.). A typical floor-plan of an OTP memory is shown in Figure 4.15. Most blocks in figure are very common to other memories (e.g. SRAM) and do not need to be explained here (see also Chapter 3). The timing circuit can be asynchronous or synchronous with an external clock. In the former case, a suitable Address Transition Detection (ATD) circuit is employed.

Let us focus on the AF bit-cell array and to show how to build up its floor-plan by considering the use of a 1.5T AF bit-cell with split gate. This cell shows a favourable ratio area-performance. The layout of the bit-cell could be that one shown in Figure 4.16.

The layer named "HV" defines the region where the oxide is thick, while WL and VPP lines are implemented with polysilicon. The AF itself is a small square of active area where the oxide under the poly is thin.

Two possible arrangements for the memory array are envisaged. Bit-cells can be placed side by side in a "chessboard" way or with adjacent

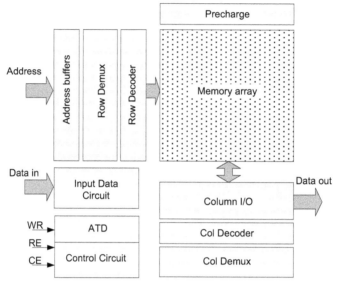

Figure 4.15   Typical floor-plan of an OTP memory.

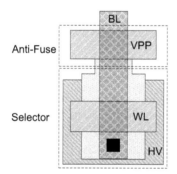

Figure 4.16   Typical layout of a 1.5T AF bit-cell.

selectors vertically flipped (see a 2 × 2 bit-cells array in Figure 4.17). The solution b) is made possible because the selector in the same column shares the same bit-line, thus saving area. The physical layout of the bit-cell proposed is easily fitted in a compact shape memory matrix. The word-lines are shared horizontally alongside the cells, while the bit-lines run vertically. The horizontal bus for the programming voltage VPP can then be collected at the ends of the array by large metal connections.

As concerns bit-cell sensitivity to the Total Dose Ionization (TID), even if the consequence of radiation influences on such breakdown channels is not

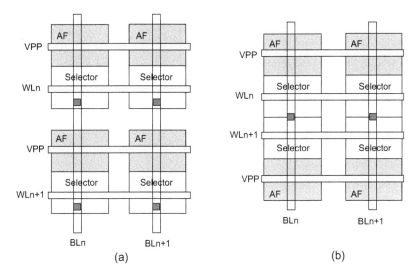

Figure 4.17 Example of 2 × 2 OTP bit-cells in a memory array: (a) "chessboard" way, (b) selectors vertically flipped.

completely clear, then the AF element turns out to be intrinsically resistant. On the contrary, the nMOS used as a selector may be sensitive to the total dose and be subject to leakage currents caused by parasitic transistor between drain-source induced by radiations.

This is particularly dangerous in case we should read a virgin (un-programmed) AF cell (see Figure 4.18), let us say corresponding to a logic "0".

When we select for reading the virgin cell (no oxide breakdown) addressed by WL_n e BL_n, the total current I_{tot} flowing across the bit-line is equal to the sum of the wanted I_{virg} current plus all the unwanted currents I_{leak} leaking through the selectors of the other cells (e.g. WL_{n+1}) because of the parasitic radiation induced MOS. Since I_{virg} is low, the sum of the I_{leak} currents can make I_{tot} high enough to be interpreted as a logic "1" instead of "0".

A second effect related to the total dose is the so-called inter-device leakage, or the generation of a conductive path caused by an accumulation of charges in the silicon oxide insulation (i.e. STI, Shallow Trench Insulation), as shown in Figure 4.19. Inter-device leakage occurs between different nMOS supposed to be insulated from each other.

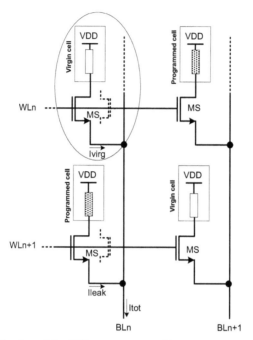

Figure 4.18   Section of the OTP array prone to TID intra-device leakage currents.

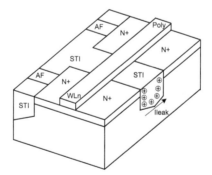

Figure 4.19   Simplified cross section of two 1.5T bit-cells belonging to adjacent word-lines (the programming line is not shown).

This has an impact on the physical layout of the memory array, since we shall guarantee that each bit-cell is isolated from the neighbourhoods. Depending on the array arrangement (a or b as in Figure 4.17), we have different parasitic leakage currents induced by radiation between bit-cells, which flow from higher potential to lower potential (Figure 4.20).

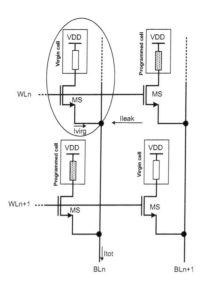

Figure 4.20 4 × 4 OTP array prone to TID inter-device leakage currents during reading.

The parasitic current I_{leak} coming from the adjacent nMOS selector can alter the correct reading of the current on the bit-cell selected via WL_n and BL_n.

The cell must also be as resilient as possible to the individual events (SEE), and in particular to latch-up phenomena (SEL, Single-Event Latch-up). Within the memory array, since there are only nMOS, any energetic ions can trigger this phenomenon. Only on the edges of the array, where it faces the decoding circuitry, which includes also pMOS, SEL could be an issue and suitable provisions shall be taken.

The peripheral circuits (decoders, demuxes, etc.) suffer from the usual effects induced by TID and SEE on digital blocks. The dose makes the transfer and timing characteristics of the combinatorial digital cells change, while the sequential logic is prone to upset that could flip the stored data. For such blocks, SEL is always an issue.

Sense amplifier senses bit-cell current and makes a comparison with a reference voltage or the current of an un-programmed bit-cell. Then, it latches the bit content. Generally, it is based on differential pairs (with an opamp input stage style) or current mirrors and is a typical analog circuit. TID causes variations in transistors parameters, i.e. increase of leakage currents, increase or decrease of threshold voltages and decrease of transconductance, g_m. The latter consequences could have a strong impact on analog circuits performance, since the accuracy of analog functions depends on transconductance and output resistance of MOS. Moreover, in analog circuits, a transient pulse can be observed, but quickly subsides (SET). Latch-up can destroy analog circuit as happens for digital. Switches which are widely used across overall OTP memory array suffer from increase of leakage currents, increase or decrease of threshold voltages and an impact on memory performance (see Section 4.3.3).

4.3.2 State of the Art

Very few OTP memories resistant to radiations exist on the market [30–32]. Generally, they are manufactured using radiation-hardened CMOS processes (adopting the Radiation-Hardening-By-Process, RHBP, approach), which is known to have higher cost, complex maintenance and hard portability. These components can sustain TID of hundreds of krad (Si), LET of some tents of MeV-cm^2/mg (Si) and are basically immune to latch-up.

In the past, some experiment has been done on AF material such as amorphous silicon [15]. In this paper, it was shown that the resistance of an un-programmed α-Si anti-fuse increases, i.e. the resistance characteristics improve with increasing total dose. The resistance of a fully programmed fuse is insensitive to ionizing radiation. On the contrary for an anti-fuse, which was partially programmed, ionizing radiation can increase the resistance of the anti-fuse sufficiently to cause a bit flip.

As said, gate-oxide AF is an excellent choice for OTP memory cell. However, some academic works explore more "exotic" solutions. The first one [33] is based on a cantilever device as a non-volatile bit-cell. The cantilever element can be used as a "one-time" as well as a "multi-time" programmable device. The cantilever is implemented in the back-end part of the standard CMOS process. Fundamental to these characteristics is the "mechanical" character of the non-volatile storage. It is based on storing a single bit of information through the mechanical state of the cantilever device. In addition, as there is no charge to leak away, the storage is completely temperature- and radiation-insensitive. Programming of the cantilever device can be done at the process's native voltage level, whatever that voltage might be: initially the cantilever switch is open. Increasing voltage results in force on cantilever. Above threshold voltage, cantilever switch closes and while removing external force, the switch remains closed (Figure 4.21). Obviously, building cantilever is an option to the standard process.

The second proposed solution [34] makes use of tantalum nitride micro-beams as anti-fuse OTP. It needs a single mask process and can be integrated

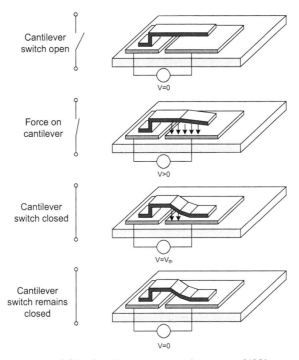

Figure 4.21 Cantilever programming event of [33].

above an integrated circuit, as back-end processes. The principle is like the previous one. The anti-fuse geometry proposed in this paper consists of a two terminals structure: one movable beam and one fixed electrode (Figure 4.22). Upon application of a DC voltage to the movable beam, it deflects and eventually gets fused to the fixed electrode. Memory reading is achieved by probing the conductance between the two electrodes, having an

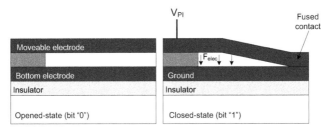

Figure 4.22 Cross-section principle of the Tantalum-Nitride OTP in the opened and closed states [34].

ideally large ratio between the original opened state (bit 0) and the fused state (bit 1). This means that one-mask process shall be integrated on bulk silicon wafers. A sweep DC bias is applied to the beam, whereas the bottom electrode is grounded. At threshold, the beam will collapse toward the fixed electrode and shorts the two electrodes. Once fused, the beam will not go back to its idle position and remains permanently fused to the bottom electrode.

Also, the Tantalum-Nitride AF is based on mechanical phenomena, so it is suitable particularly for rugged environments (high temperature/radiation or vibration).

The solutions presented above to make the OTP bit-cell robust against radiation share the disadvantage to be incompatible with standard CMOS technologies, because they need at least one more step very specialized. Thus, their use is limited to stand-alone component, and cannot be embedded in complex SoC technology. However, in space as well as in the consumer market, as the shrinking of the electronic components enables it, the trend is towards more dense and integrated systems, aiming at continuous reduction of volume, mass, and power consumption. This is the reason why adopting an RHBD methodology on standard CMOS technology is convenient. The approach used is the RHBD techniques to improve the total ionizing dose (TID) hardness, avoid the latch-up of the component and reduce the effects of the high-energy crossing particles.

4.3.3 RHBD Mitigation: Architectures and Layout

Even for OTP it is advisable to apply the well-known RHBD methodology, which can be divided into three levels: architectural, circuital and layout design. Each of the three levels is equally important and addresses different solutions to the radiation effects. For the OTP two different sections have to be considered, the periphery and the bit-cell. The RHBD design methods used for the design of a generic OTP memory are exposed in the following. Even though those related to the periphery blocks (decoder, demux, etc.) have been discussed in detail in Chapter 2, they will be briefly repeated here at the reader convenience.

When designing the logic gates employed in the periphery macro-blocks, the main rules followed at circuit level to generate a rad-hard standard library can be resumed as [35]:

- Reducing the number of floating nodes using minimal number of transistors not directly connected to power lines to reduce SEU and latch-up;

- Avoid the use of pass transistors when possible;
- Increase the parasitic capacitance of critical nodes to reduce the probability that an SET becomes an SEU;
- Redundancy of decision nodes to reduce MBU (Multiple-Bit Upset) due to address upsets.

At layout level, the SEU, SET, TID and latch-up aspects are addressed. First, to reduce the TID, edge-less transistors are used in the periphery of the arrays to avoid the effects related to the charge collected in thick oxides. SEU and SET events are analyzed for critical cells and special precautions are taken. For example, critical charge in sequential logic ports is increased by the addition of capacitors. Finally, to avoid the latch-up, enhanced guard rings surround all pMOS and nMOS transistors.

As concerns specifically the proposed AF OTP bit-cell, the intra-device leakage can be mitigated by using layout solutions mentioned in previous chapters, i.e. by adopting MOS selector with either higher length, L, or suitable physical shapes such as Edge-Less Transistor (ELT) or Dog-Bones (DB). Figure 4.23 show three possible selector shapes for the bit-cell, where the ELT is clearly the biggest, even because the annular shape does not allow to implement low W/L ratio. The proportion between shapes shown in the figure is almost realistic for an 180 nm standard CMOS technology. It comes out that DB bit-cell is about 10% bigger than linear one, while ELT occupies double area. It is expected that DB shape could tolerate 100 krad

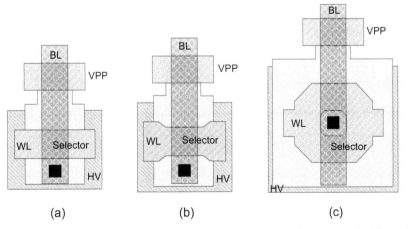

Figure 4.23 Physical layout of an AF OTP bit-cell with selector implemented with regular shape (a), Dog-Bone shape (b) and ELT shape (c).

TID, while ELT could support up to 300 krad. The choice of the most suitable shape is the result of a trade-off between consumption and insensitivity to radiations.

The intra-device leakage can be tackled by increasing the isolation between devices. In Figure 4.24 the possible leakage between the two bit-cells belonging to two consecutive WL must be avoided (selector is designed with linear transistor). To overcome this phenomenon, the more drastic solution provides the insertion of a substrate bias line connected to ground in between the two bit-cells in both vertical and horizontal directions, often referenced as EGR (Figure 4.25). A less area-consuming solution foresees the insertion of some equally spaced polarization taps, often referenced as IGR (see Chapter 3).

Figure 4.26 presents an example of a real layout of a 4×4 array of AF bit-cells implemented with a standard 180nm technology. The array on the left-hand side is done with selector having regular shape, while those in the middle and on the right-hand side with DB and ELT, respectively. The different occupied area is noticeable. Table 4.2 reports the normalized area ratios. In order not to waste further area, substrate guard-rings have been inserted only horizontally, one every two bit-cells rows (in red rectangles). This is not an RHBD mitigation only but also a precaution to guarantee a correct ground distribution.

The area increment of the 4×4 array with AF having ELT selector is due to bigger size of the bit-cell itself (see Figure 4.23) and to the fact that with ELT-shape it is impossible to set the bit-line connection in common between two adjacent flipped bit-cells (Figure 4.27). The nodes (A and B) in common between AF and selector MOS cannot be shared by the two bit-cell.

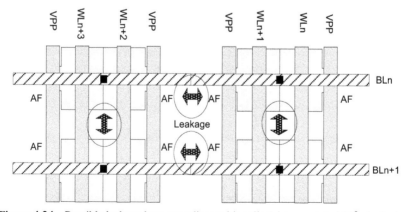

Figure 4.24   Possible leakage between adjacent bit-cells (picture rotated 90° clockwise).

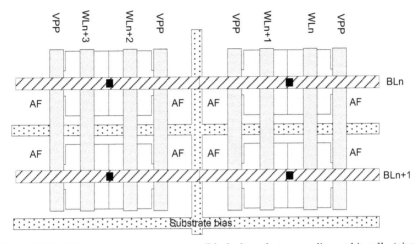

Figure 4.25 Substrate taps to prevent possible leakage between adjacent bit-cells (picture rotated 90° clockwise).

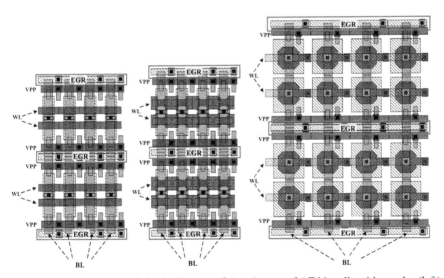

Figure 4.26 Examples of physical layout of 4 × 4 array of AF bit-cells with regular (left), DB (middle) and ELT (right) selector shapes.

Table 4.2 Normalized area ratio for a 4 × 4 AF 1.5T array with different bit-cell shapes

	1.5T-Lineare	1.5T-DB	1.5T-ELT
Area (norm.)	1	1.1	2.5

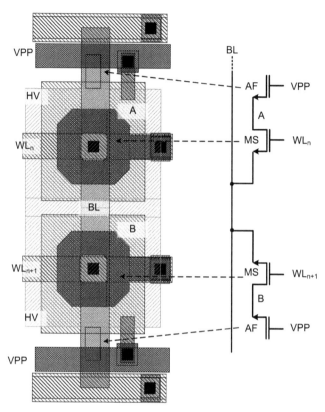

Figure 4.27   Two bit-cells (AF + selector with ELT shape) placed side-by-side and vertically flipped.

At architecture level, the organization of the array can be that presented in [35], where the whole memory array is divided into eight tiles each one corresponding to 1 bit of the 8 bit word-data out. Row and column decoders are repeated for each tile in such a way that a failure on a row or a column will produce an error on one bit only of the output byte to make it easier EDAC operation.

Moreover, another important issue to be considered is how the bit-cell content is sensed. The typical working principle of the OTP foresees the use of a bit-cell and a voltage/current reference, or of two bit-cells, the storing cell and a reference cell. When the bit-cells are exposed for long periods of time to radiation the comparison of the storing cell with the reference bit-cell could become problematic. It is also important to notice that sometimes a single reference is used for a whole memory, which may limit the radiation immunity of corresponding devices.

To increase the radiation tolerance of a non-volatile memory, Ref. [36] proposes to use a differential bit-cell instead of using a single critical reference cell. In this case, two bit-cells are used to store each bit, one of the cells is programmed while the second one is left in the un-programmed state. In Figure 4.28, a simplified view of the differential architecture is represented. On the left-hand side, the bit array is placed, and on the right-hand side, the negated bit array is placed. The programming of the array is done byte-a-byte, selecting the left- and right-hand sides and programming the cells depending on the input bit. For instance, if a logic "1" has to be stored in the dark gray bit-cells, the left cell has a low R_{ON} (broken oxide) and the right cell has a high R_{OFF} (virgin oxide). On the contrary, if the light gray bit-cells are programmed with a logic "0", the left cell is left un-programmed (virgin), while right cell is programmed (broken). Then, during the reading of the memory, the two bit-cells are compared, and the data out become available. The differential architecture increases the operation window, since we do

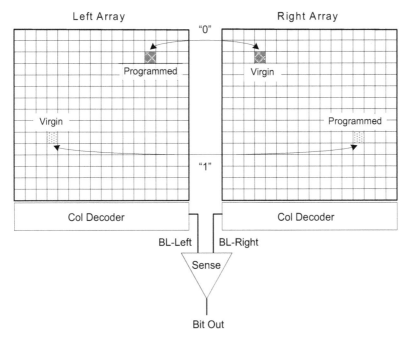

Figure 4.28 The two memory cells composing a bit are separated into two independent arrays to avoid the effects of single events.

not have the need of a precise reference. Moreover, we increase also the sense amplifier robustness since it works with a double input signal. This is particularly important for OTP AF bit-cell based on the oxide breakdown, where the sense margin is a critical parameter and depends strongly on how the oxide is broken. However, this method reduces the storing density.

Also, with respect to SEE, this architecture is robust. Studies have shown that in deep sub-micron technologies, the cross of a high-energy particle can cause a loss of charge in a cloud of memory cells. The size of the cloud depends not only on the energy and angle of the particle, but also on the technology used. Considering this possibility, the division into two independent arrays, enlarging the distance between the two cells of the single bit, reduces the probability of losing the information due to an SEE (see Chapter 3 for further details).

The problem related to the leakage currents induced by TID does not affect only the selector MOS, but also all the transistors used as switches in the reading and writing paths (see, for example, Figure 4.29). These parasitic currents can alter the reading current of a logic "0" (corresponding to virgin cell) and reduce the read window. This issue can be considered a typical issue related to an analog design, where the sizes of switches are always the results of a trade-off. Two main phenomena shall be considered related to switches [37]:

- The R_{on} of the CMOS switch shall allow enough speed;
- The clock feed-through shall be kept at minimum.

The first condition pushes for using complementary switches with higher aspect ratio, while the clock feed-through pulls for minimum MOS dimensions. However, ELT topology imposes constraints in the choice of MOS W/L. In particular, the annular shape fixes a limit on the minimum width, which is possible to draw, so that the switches size does not correspond to minimum lithographic. In the particular case of the OTP reading scheme in Figure 4.29, we introduce a time constant related to the non-zero resistance of switches R_{on}, the resistance of bit-cell R_v and bit-line capacitance C_{bl} and sense input capacitance that can increase memory access time during reading:

$$\tau_c = (R_v + 3R_{on}) * (C_{bl} + C_{in})$$

The sense amplifier is the other circuit that can be related to analog. The differential stage with common source is commonly used to detect differential bit-cell content. Before amplification the current coming from the bit-cell, I_{cell} (and its counterpart I_{cell_n}) through bit-lines BL-Left and BL-Right

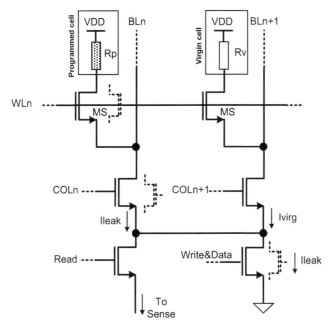

Figure 4.29 Possible effect of switches leakage due to TID in adjacent bit-cells during reading the virgin cell.

shall be transformed into a voltage by means of an I/V circuit, e.g. a diode connected MOS (Figure 4.30).

As happens in SRAM, the inputs of the differential pair are connected by the bit-lines to a large number of cells, whose distributed capacitance can be significant; for this reason, if the OTP memory size is big, bit-lines could be pre-charged to make the sense amplifier work in best conditions. The lower nMOS transistor enables the sensing operation (EN).

Even if they are more prone to TID, input nMOS transistors have been used in the differential pair because of the dynamic of the signals to be compared. Two main performances are impacted by the dose:

- open-loop gain and
- input offset voltage.

The open-loop gain of the differential stage is always given by the ratio of transconductances and conductances. Since radiations cause variations in transistors parameters, i.e. increase or decrease of threshold voltages, decrease of transconductance, g_m, and transistors output resistance, the gain

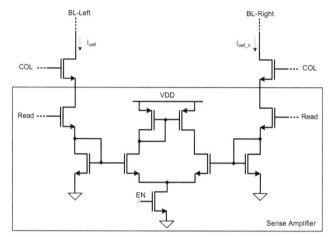

Figure 4.30   Schematic of a sense amplifier for Anti-Fuse bit-cell.

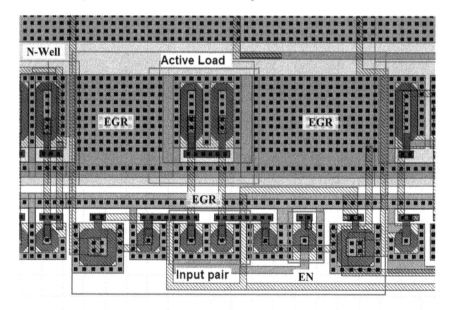

Figure 4.31   Layout of a sense amplifier (blue rectangle) for Anti-Fuse bit-cell.

could change. To avoid this, it is possible to utilize all ELT shapes and design the stage with a sufficiently high gain in normal conditions. The layout of the sense amplifier in Figure 4.30 is shown in Figure 4.31. Extended guard-rings can be noted for both substrate and n-well to separate pMOS from nMOS transistors.

Besides usual causes of offset, such as electrical biasing differences, lithographic effects and mechanical stress, radiations induce complex input offset voltage drifts. Layout precautions like centroid input differential pair and symmetric architecture can reduce the conventional input offset voltage. These techniques are common practice when dealing with regular shape transistors, where they need matching. However, this practice is not straight-forward with ELT transistors. In order to accomplish the matching, it is possible to apply a variant of the interdigitated MOS to ELT, for example, on the MOS of input pair. They can be split into n ELT, each of them placed in an "interdigitated" way. Figure 4.32 shows a layout example of interdigitated differential pair consisting of M1 and M2 input pair, together with dummy transistors (MD) to make the edges of internal pMOS homogeneous.

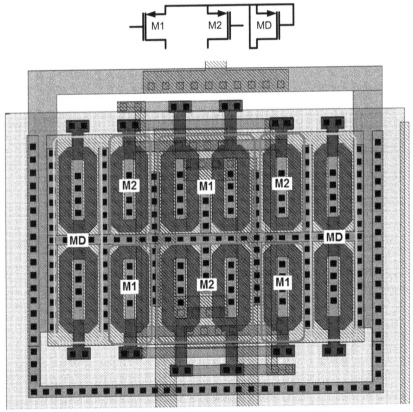

Figure 4.32 Layout example of interdigitated differential pair and dummy pMOS MD.

4.4 Conclusion

In this chapter we have analyzed the basic concept of OTP Memories and how it is possible to use those based on anti-fuse in space applications. This target is accomplished thanks to the intrinsic properties of such memories and through the application of special design techniques, i.e. Radiation-Hardened-By-Design, to standard CMOS technologies, which make them resilient to radiation effects. Even if rad-hard OTP Memories are yet in an early development stage, the expected results are promising and will represent a technological breakthrough for on-board components and an added value to space microelectronic industry.

References

[1] L. Hong, "Comparison of Embedded Non-Volatile Memory Technologies and Their Applications", White Paper Kilopass, May 2009.

[2] A. K. Sharma, "Advanced Semiconductor Memories - Architectures, Designs, and Applications", Wiley, 2003.

[3] J. Peng, G. Rosendale, M. Fliesler, "A novel embedded OTP NVM using standard foundry CMOS logic technology", NVSMW, 2006.

[4] Y. Bentoutou, "Program Memories Error Detection and Correction On-Board Earth Observation Satellites", International Journal of Electrical, Computer, Energetic, Electronic and Communication Engineering Vol. 4, N. 6, 2010, pp. 933–936.

[5] H. B. Riblet, "The Small Astronomy Satellite-3 - A General Description", APL Technical Digest, Vol. 14, N. 3, 1975, pp. 1–5.

[6] A. Arbat, C. Calligaro, V. Dayan, E. Pikhay, and Y. Roizin, "NVM for FPGA booting in space", Adaptive Hardware and Systems (AHS), NASA/ESA Conf., June 2013.

[7] E. Ebrard, B. Allard, P. Candelier, P. Waltz, "Review of fuse and antifuse solutions for advanced standard CMOS technologies", Microelectronics Journal 40 (2009), pp. 1755–1765.

[8] J.-K. Wee, et al., "Anti-fuse Circuits and their Applications to Post-Package of DRAMs", Journal of Semicond. Techn. and Science, Vol. 1, N. 4, Dec. 2001.

[9] O. Kim, C. J. Oh, K. S. Kim, "CMOS trimming circuit based on polysilicon fuse", Electronics Letters 34, 1998, pp. 355–356.

[10] M. Alavi, et al., "A PROM element based on salicide agglomeration of polyfuses in a CMOS logic process", Int. Electron Devices Meeting, IEDM 1997, pp. 855–858.

[11] S. H. Kulkarni, et al., "A 4 kb Metal-Fuse OTP-ROM Macro Featuring a 2 V Programmable 1.37 um^2 1T1R Bit Cell in 32 nm High-k Metal-Gate CMOS", IEEE J. Solid-State Circ., Vol. 45, N. 4, April 2010, pp. 863–868.

[12] S. H. Kulkarni, et al., "High-Density Metal-Fuse Technology Featuring a 1.6 V Programmable Low-Voltage Bit Cell With Integrated 1 V Charge Pumps in 22 nm Tri-Gate CMOS", IEEE J. Solid-State Circ., Vol. 51, N. 4, April 2016, pp. 1003–1008.

[13] W. Kurjanowicz, "Evaluating Embedded Non-Volatile Memory for 65 nm and Beyond", Sidense, DesignCon 2008.

[14] J.-S. Choi, et al., " Antifuse EPROM Circuit for Field Programmable DRAM", 2000 IEEE International Solid-State Circuits Conference, Feb. 2000, pp. 406–407.

[15] J. M. Benedetto and C. C. Hafer, "Ionizing Radiation Response of an Amorphous Silicon Based Antifuse", Radiation Effects Data Workshop, 1997.

[16] C. de Graaf, et al., "A Novel High-Density Low-Cost Diode Programmable Read Only Memory", International Electron Devices Meeting, IEDM 1996, pp. 189–192.

[17] P. Candelier, N. Villani, J. P. Schoellkopf, et al., "One time programmable drift antifuse cell reliability", Proceedings 38th Annual 2000 IEEE International Reliability Physics Symposium, 2000, pp. 169–173.

[18] Patent N. US 7,755,162 B2, Kurjanowicz et al. (Sidense), "ANTI-FUSE MEMORY CELL", Jul. 13, 2010.

[19] Sidense Web page: "Technology" (https://www.sidense.com/technology).

[20] R. Barsatan, M. T. Yin, C. Mansun, "A zero-mask one-time programmable memory array for RFID applications", Proceedings IEEE International Symposium on Circuits and Systems, 2006, pp. 975–978.

[21] N. D. Phan, I. J. Chang, and J.-W. Lee, "A 2-Kb One-Time Programmable Memory for UHF Passive RFID Tag IC in a Standard 0.18 m CMOS Process", IEEE Trans. on Circuits and Syst. - I, Vol. 60, N. 7, July 2013, pp. 1810–1822.

[22] J. Kim, K. Lee, "Three-transistor one-time programmable (OTP) ROM cell array using standard CMOS gate oxide antifuse", IEEE Electron Device Lett, 2003, 24, pp. 589–591.

[23] H.-K. Cha, et al., "A 32-KB Standard CMOS Antifuse One-Time Programmable ROM Embedded in a 16-bit Microcontroller", IEEE Journ. of Solid-State Circuits, Vol. 41, N. 9, Sept. 2006, pp. 2115–2124.

[24] H. Ito, T. Namekawa, "Pure CMOS One-time Programmable Memory using Gate-Ox Anti-fuse", IEEE 2004 Custom Integrated Circuit Conf., pp. 469–472.

[25] L. Xian, Z. Huicai, J. Cheng, and L. Xin, "A 4-kbit low-cost antifuse one-time programmable memory macro for embedded applications", Journal of Semiconductors, Vol. 35, No. 5, May 2014.

[26] X. Li, H. Zhong, Z. Tang, and C. Jia, "Reliable Antifuse One-Time-Programmable Scheme With Charge Pump for Postpackage Repair of DRAM", IEEE Trans. on Very Large Scale Integration (VLSI) Syst., Vol. 23, N. 9, Sept. 2015, pp. 1956–1960.

[27] T. Namekawa, et al., "A 65 nm Pure CMOS One-time Programmable Memory Using a Two-Port Antifuse Cell Implemented in a Matrix Structure", IEEE Asian Solid-State Circuits Conference 2007, pp. 212–215.

[28] S.-Y. Kim, J.-H. Lee, T.-Y. Kim, Y. You, "Design of CMOS Dual Antifuse OTP Memory Based On Gate Oxide", SoC Design Conference (ISOCC), 2014 International, pp. 284–285.

[29] H.-K. Cha, I. Yun, J. Kim, B.-C. So, K. Chun, I. Nam, K. Lee, "A 32-kB standard CMOS antifuse one-time programmable ROM embedded in a 16-bit microcontroller", IEEE Journal of Solid State Circuits, Vol. 41 (2006), pp. 2115–2124.

[30] Datasheet "UT28F256LVQLE Radiation-Hardened 32K x 8 PROM", Cobham, March 2007.

[31] Datasheet "Radiation Hardened 8kx8 CMOS PROM", Intersil, Aug. 2000.

[32] Datasheet "Programmable read-only memory 32K x 8 radiation-hardened PROM — low power 3.3V", BAE SYSTEMS, 2016.

[33] M. A. Beunder et al., "New Embedded NVM Technology for Low-Power, High Temperature, Rad-Hard Applications", Non-Volatile Memory Technology Symposium, 2005, pp. 65–68.

[34] P. Singh, C. G. Li, P. Pitchappa, and C. Lee, "Tantalum-Nitride Antifuse Electromechanical OTP for Embedded Memory Applications", IEEE Electron Device Letters, Vol. 34, No. 8, Aug. 2013, pp. 987–989.

[35] A. Arbat, Member, IEEE, C. Calligaro, Y. Roizin, and D. Nahmad, "Radiation Hardened 2Mbit SRAM in 180nm CMOS Technology", Satellite Telecommunications (ESTEL), 2012 IEEE First AESS European Conference on.

[36] A. Arbat, C. Calligaro, V. Dayan, E. Pikhay, Y. Roizin, "SkyFlash EC Project: Architecture for a 1Mbit S-Flash for Space Applications", Electronics, Circuits and Systems (ICECS), 2012 19th IEEE International Conference on.

[37] U. Gatti, C. Calligaro, E. Pikhay; Y. Roizin, "Radiation-hardened methodologies for CMOS flash ADC", Analog Integrated Circuits and Signal Processing, Vol. 84, N. 3, Sept. 2015, pp. 141–154.

5

Rad-hard Flash Memories

Anna Arbat Casas

XGLab S.R.L., Milano, Italy

"Knowing where the trap is—that's the first step in evading it."
Frank Herbert, Dune

The Flash non-volatile Memory cell operation principle is based on the change of the current flowing through a CMOS transistor due to the charge of the transistor threshold produced in the programming stage. The effect of the radiation is the modification of the stored data due to the interaction of the radiation with the trapped charge, producing a shift of the memory cell transistor threshold. In this chapter, we will focus our attention on how to reduce the effects of the radiation considering the architecture point of view. We will discuss how to apply the radiation hardening techniques, previously proposed, to this particular structure. The use of a differential cell will be presented, replacing the classical single-reference cell architecture of flash memories. Finally, we will also introduce the requirements for some auxiliary circuits, such as charge pumps and pad rings.

5.1 Introduction

In this chapter, we will focus on a different type of reprogrammable non-volatile memories: Flash memories. Two chapters are devoted to this type of memories: in the first one, the radiation enhancement considerations from the architecture point of view are presented, while in the second, they are discussed from the technology perspective.

In the first part of the chapter, an overview of the typical reprogrammable Non-Volatile Memories (NVM) is given. Then, the principle of operation of the Flash memories is exposed, which are the focus of the chapter. The difference between the NAND and NOR Flash type is shown in the last part of the first section.

In the second part of the chapter, the effect of the radiation on Flash memories and an exhaustive study of the RHBD methodology applied to improve the radiation hardening in this type of memories are presented. The study begins from the single cell evaluation to the general memory architecture. Finally, in this section, a case study where these techniques have been used for a particular type of Flash cell is presented.

In the third part, the study considers two side structures of the memory: the charge pumps and the pad ring. As in the memory architecture, general considerations will be outlined without analyzing a particular circuit. Also, in this section, the side circuits used in the case study presented in the previous section are shown.

5.1.1 Non-volatile Memories Overview

Two main types of NVM using semiconductors technology are available in the market: ROM-based memories (PROM, EPROM and EEPROM) and non-volatile-RAM-based memories (ferroelectric RAM, magnetoresistive RAM, phase-change RAM among others) [1]. Flash memories – the aim of this chapter – are an evolution of the EEPROM memories that can be read in a random access, so technically, flash memories can be categorized in both types.

The read-only memory (ROM) is a type of non-volatile memory developed to contain fixed information, which cannot be changed after manufacture. Early ROM memories were diode matrix or mask ROM, which were physically encoded during the fabrication of the component. To include some degree of modification and to increase the flexibility reducing fabrication cost, ROM manufacturers proposed programmable read-only memories (PROM), or one-time programmable ROM (OTP), which arrive to the user as virgin devices and can only be programmed once: high voltage is used to create or destroy permanently an internal link (i.e. fuses/anti-fuses) of the chip. Data retention for those memories is not affected by the media, as the information is based on a physical state and not on an electrical permanence of the integrated circuit (IC).

The requirement to change this information while storing it over time produced the evolution of those memories to erasable-programmable devices. This was obtained with the introduction of the floating-gate transistor structure, such as the erasable programmable read-only memories (EPROM) and the electrically erasable programmable read-only memories (EEPROM).

EPROMs and EEPROMs consist of an array of transistors with a gate terminal protected with a high-quality insulator. To program a cell, a high voltage is applied to the gate terminal – switch – producing a trapping of electrons on the far side of the insulator. The accumulation of charge switches the transistor to the "on" state ("1"). The accumulated charge can be removed by using UV light in the case of EPROMs or by applying a pulsed high power through the source – drain terminals in the case of EEPROMs. In both cases, erase and programming are slow actions requiring long time and degrading the insulator of the memory cell, limiting the erase/write cycles' endurance to values between 1,000 and 1,000,000. Data retention may be limited by charge leaking from the floating gate, increased by high temperatures or radiation.

Both types of Flash memories (NAND [2] and NOR [3]) were developed in the early 1980s as an evolution of the EEPROM.

Flash memory is based in the floating-gate MOSFET structure. The floating gate acts as a trapping media for the electrons forced during writing of the cell, modifying the threshold voltage of the MOSFET transistor, and so changing the drain current during reading operation. As in EEPROM, the electrons can be released from the floating gate by applying an electric field, and so stressing the oxide during the write/erase cycles. In the case of flash cells, the density of the memory can be increased by using the multi-bit techniques. In this case, it is possible to store more than one bit of information per cell setting different threshold levels corresponding, for example, to "00", "01", "10" and "11" states [4]. This strategy increases the storing density, thus in turn increasing the complexity of the programming and reading cycle.

Newer approaches to obtain non-volatile memories are based on RAM structures, such as ferroelectric RAM (FRAM), magnetoresistive RAM (MRAM) and phase-change RAM (PCM) among others [5]. The main difference between those structures and the typical RAM structure consists in the fact that the information is not stored as charge in a cell that will be lost when power is turned off, but in a modification of a physical parameter of the cell, which is kept event when the power is turned off. In the case of FRAM instead of the dielectric layer of the DRAM, it contains a thin ferroelectric film, which changes the polarity in an electric field, producing a binary switch. FRAM has a higher endurance than other NVM devices, as well as high write speed and gamma radiation tolerance. MRAM uses a magnetic element inserted in the RAM cell, while PRAM reading method is based on the change of the electrical resistance.

5.1.2 Flash NVM: Principle of Operation

The Flash memory is an array of memory cells based on the floating-gate MOSFET structure presented in Figure 5.1. The structure contains two gates instead of one. The top gate is the control gate (CG) and the bottom gate is the floating gate (FG). The floating gate can be fabricated using polysilicon (conductive gate) or nitride (non-conductive gate). The FG is insulated with an oxide layer and behaves as a trapping medium for the electrons forced there during programming of the cell. Once the FG is charged, the electrons partially reduce the electric field from the CG, increasing the threshold voltage (V_{T1}) of the cell. To read the cell, an intermediate voltage is applied to the CG. When the channel conducts, the FG is not charged, and when the channel does not conduct, the FG is charged. Each state corresponds to a logic value. So, the presence of a logical "0" or "1" is sensed by determining whether current is flowing or not through the transistor when an intermediate voltage is applied to the CG.

In the case of multilevel cell, different current flows are sensed in order to determine the charge level in the FG.

Two different configurations of the flash memories are available: NOR-flash and NAND-flash. The specific details for these two configurations will be detailed in the following section, while a general overview is exposed here.

Nowadays, flash memories contain a single supply voltage. To obtain the different internal voltages required for programming and erasing the memory on-chip, charge pumps are used. This is one of the most sensible components when developing a flash memory for high-radiation environment. This means that writing/erasing of the memory may not be operative after a certain period

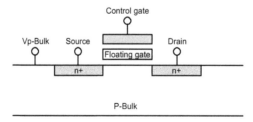

Figure 5.1   Cross section of a generic floating-gate Flash memory cell.

due to radiation damage, while reading will continue to be operative for a longer period.

The limitations of the flash memories are:

- **Block erasure:** although read/write of the memory can be done in a single bit random access, the erase is done by blocks. Erasure sets all the bits of a block to a "1", and then it is possible to set each single bit to a "0" independently during programming.
- **Limited number of program/erase cycles:** as in the case of EPROMs and EEPROMs, also flash cells program and erase cycles produce the passage of electrons through the insulator, which is mechanically stressed with this procedure. This limits the number of program/erase cycles of each individual block of the memory. The division of the memory in independent erasable blocks allows reducing the damage by controlling the blocks containing information. For this reason, flash memories require a block memory management to ensure a longer lifetime of the memories even though part of the memory may not be useful anymore.
- **Read disturb:** the read disturb is the effect that produces the modification of the nearby memory cells while reading. This requires many read operations of the same cell to modify the surrounding cells. Damage can be avoided by controlling the total amount of read operations of a block, transferring the information to a new block and erasing the old one to free the space.
- **X-ray erase:** the effect of the X-ray is to erase the programmed cells of a flash memory.

5.1.3 Flash NVM Standard Architecture Overview: NOR vs. NAND

The basic cell of the flash memory is an NMOS transistor with a floating gate as a trapping media. Based on this cell, two different architectures are available for the flash cells: NOR-Flash and NAND-Flash. In NOR-flash, the connections of the cells are made in parallel as the NMOS transistors in the NOR gates (all cells have a ground connection on one terminal). In NAND-flash, several transistors are connected in series to the ground connection as in the NAND gates.

5.1.3.1 NOR-Flash

Each single memory cell contains the drain of the transistor connected to the ground and the source connected to the bit-line. The control gate is connected to the word-line. Figure 5.2 shows the general connection for NOR-flash memory. When a word-line is selected (high voltage), the memory transistor pulls the bit-line low (the current flow depends on the FG trapped charge, and thus on the threshold voltage). NOR-flash memories are read and program in random access, and erase is done in blocks. Programming of the memory changes bits from a logical one to a zero. So, bits that are already zero are left unchanged. Erase of the memory sets all the bits of a block back to one. Due to the internal structure of the memory, it is possible to implement the read-while-write functionality.

To program a single-level NOR-flash memory cell (set to "0"), it is required to set a high voltage to the CG (typically Vh > 5V), producing a current flow between the drain and source terminals of the NMOS transistor (Figure 5.3). The current is high enough to cause hot-electron injection through the insulator layer onto the FG, where the electrons are trapped.

To erase a NOR-flash (set to "1"), it is required to apply a high voltage of the opposite polarity to the CG in order to pull off the electrons of the FG through quantum tunnelling (Figure 5.4). Erase operation is performed in segments of memory.

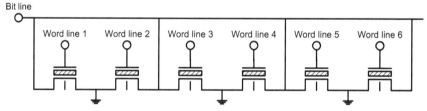

Figure 5.2   NOR-Flash memory connection architecture.

Programming via hot electron injection

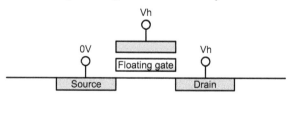

Figure 5.3   Programming connections of the Flash cell terminals.

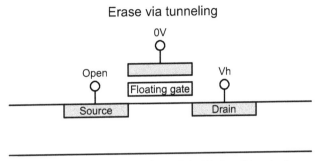

Figure 5.4 Erase connections of the Flash cell terminals.

5.1.3.2 NAND-Flash

In the case of the NAND-Flash, several transistors are connected in series between ground and a bit-line and controlled with different word-lines (Figure 5.5). The bit-line is pulled low only if all the word-lines are pulled high (word-line voltage above threshold voltage). To read a cell, the bit-line group is selected, the rest of the word-lines are pulled up above the programming threshold voltage, while the selected cell is pulled up to a value below the programmed threshold voltage. The selected group will conduct only if the cell is not programmed. The NAND-flash structure increases the addressing complexity of the memory, as well as the density. NAND-Flash are written using tunnel injection of electrons and erased using tunnel release. NAND-Flash memories are read and program on page-wise, and erase in block-wise. NAND-Flash memories require Error Correcting Code (ECC) checksum circuits. To increase the reading speed and to obtain the advantages of random-access memories, virtual memories strategies are often used, adding to the system an external RAM and requiring a memory management unit (MMU).

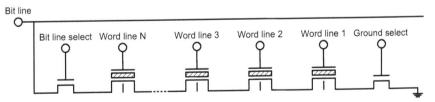

Figure 5.5 NAND-Flash memory connection architecture.

5.1.3.3 Summary

The distinctions between NOR and NAND flash are:

- The connections of the individual memory cells.
- The interface provided for reading and writing the memory (random for NOR, page for NAND).
- The size of NOR cells is larger than NAND cells due to the distribution of connections.
- Read speed in NOR cells is better than in NAND cells.
- NOR cells are read using low voltages, while NAND cells require the use of a higher voltage to be applied to the rest of the cells in the block.

Considering the two structures presented and the fact that the use of a high voltage in a radiation environment increases the damages of the whole IC, the best structure for a Flash memory in a radiation environment is the NOR-Flash structure.

5.2 Radiation-hard Flash Architecture Study

The effects of the radiation onto ICs can be divided into accumulative (long exposition periods) and single events (passage of a high-energy particle producing the generation of electron–hole pairs). Some of these effects can be avoided using the design guides of the RHBD techniques presented in previous chapters [6, 7].

Considering the periphery of the memory, it is important to determine the total ionizing dose (TID) effects and the single event effects (SEE). TID produces the trapping of charges, mainly positive, within the oxide. The number of charges is larger in the transition regions between the gate oxide and the field oxide. The accumulation produces a leakage current between drain and source in NMOS transistors. Edge less transistors (ELT) are used to avoid the leakage as they avoid oxide transition between drain and source [8]. SEE effects like single event transitions (SET) or single event upsets (SEU) can be reduced incrementing the node capacitance of critical nodes, such as decisional stages and flip-flops [9]. As concerns the periphery, the approach followed to improve the system radiation hardness is similar to RHBD techniques presented for the SRAMs [7].

Considering the Flash memory cell, the most important issue is the shift of the threshold voltage when irradiated. The variation of the threshold is due to the loss of charge stored in the trapping media and/or to the accumulation of positive charge which compensates the trapped electrons during programming [10, 11].

In the following part of this section, it is analyzed how to enhance the architecture of the Flash memory to improve the radiation hardness, beginning from the Flash cell itself, considering the program, erase and read operations and ending with the general bit architecture. In the last part of the section, a real example using the methods discussed here is given.

5.2.1 Critical Features: Introduction of the Differential Cell Architecture

The most critical feature of the Flash cell is the shift of the threshold voltage when exposed to irradiation for long periods (TID) [10]. The shift affects not only the programmed cells, but also the non-programmed or virgin cells. This is since not only the electrons trapped in the floating gate are released due to the radiation effect, but also that positive charge remains trapped in the gate oxide. Figure 5.6 shows the evolution of the threshold voltage of the flash cells in a 4 kbit prototype irradiated with 70 krad of gamma ray [12]. Initially, the memory was programmed with a "01" pattern, so half the bits are set at the initial state and half of them are set to a high-voltage threshold. The distribution of virgin bits (V_{th} = 2 V) is much narrower than the distribution of programmed bits (V_{th} = 4.2 V). This is due to internal limitations of the prototype, which are not part of this discussion. The shift of the cells reduces the initial window between programmed and clear cells from 1.5 V to 1 V after a TID of 70 krad in the example.

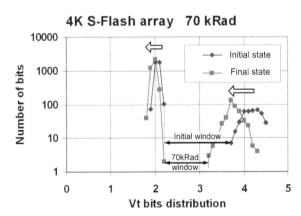

Figure 5.6 Threshold distribution of a 4 kbit S-Flash array before irradiation and after an accumulated dose of 70 krad [12]. All the cells show a drift to a lower threshold voltage, and the operation window between programmed and virgin cells is reduced.

The shift of the threshold voltage can be compensated using a control gate voltage near the low threshold. But the typical control gate voltage used in these memories is the power supply voltage. This means that using a Flash memory powered at 3.3 V, after a total dose of 70 krad, part of the memory will not contain the correct information. Of course, this can be solved by reprogramming the memory more frequently, but this reduces the lifetime of the system and increases the necessity of component replacement, which is not always possible (i.e. during long-term space missions).

Another more convenient approach can be followed to increase the periods between reprogramming of the cell, and so increasing the component lifetime. This approach consists in the use of differential flash cell architecture. In this case, to store one bit, two flash cells are used. To store the information, one of the Flash cells is programmed and the second one is kept virgin. To read the information, both control gates are powered at the same voltage and the currents flowing through them are sensed. Figure 5.7 shows the structure of the differential flash cell where two bits of information are stored. Drain lines of the flash cells are connected to the ground (D_bl0 and D_bln0). Source lines of the flash cells are connected to the bit-line (S_bl0) and to the negated bit-line (S_bln0). Finally, control gates are connected to the word-line (wl0 for the first bit and wl1 for the second bit). In the first bit (wl0), the left cell is kept virgin (bl) and the right cell is programmed (bln), which will correspond to the logical "0" information. On the second bit (wl1), the programmed cell is the bl cell, and it will correspond to the logical "1".

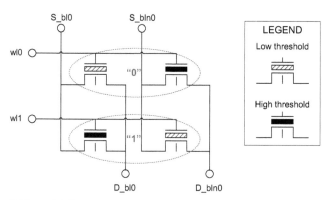

Figure 5.7   Differential Flash cell structure. The cell is composed of two basic Flash cells, one with a low threshold (virgin cell) and one with a high threshold (programmed cell). It is defined as a convention to store a bit, i.e. top differential cell contains a "0" and bottom differential cell contains a "1".

From the point of view of single events (SEE), the main problem is related to the typical size of the Flash cell, which is typically the minimum size for an NMOS transistor in the used technology. This means that, if a high-energy particle crosses the memory, the effect of the particle will reach more than one flash cell. If the two flash cells of the differential structure are placed near in the memory array, both could be affected by a single particle and the information may be lost.

A possible solution is to place the two cells in distant positions inside the array. This structure will increase the complexity in the decoding system and in the reading structure, as it will increase the difficulty to compensate the RC contribution of the different lines. Another option is to use a specular approach. In this case, two specular arrays are used: one containing the bit and one containing the negative bit information. Figure 5.8 shows the configuration of the specular memory. In this case, line decoders for reading and programming/erasing operations can be positioned on the top/bottom of the array, while the word decoder can be positioned in the middle of the two arrays. This structure ensures the RC compensation of both cells during all the operations of the memory and also of the decoding structure. The positioning of the word decoder in the centre of the two arrays reduces the error decoding possibility and increases the protection of the SEE effects, as the separation of the two Flash cells corresponding to a single bit is increased.

Using the differential cell architecture in specular independent arrays also modifies all the rest of required structures to program, read and erase the whole memory. In the following sections, each of these circuits will be analyzed separately.

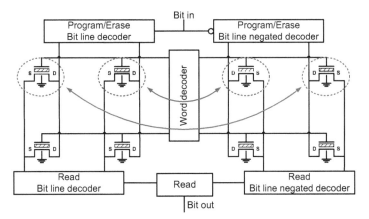

Figure 5.8 Distribution of the differential cell components in the specular symmetric array.

5.2.2 Program – Erase Circuit Analysis

When considering the different circuits of the periphery of the Flash array, it is possible to divide them into two groups, those requiring high voltages (program and erase) and those using low voltages (read and verify). The first consideration to enhance the radiation hardness of this memories is to separate the circuits working at high voltages from those operating at low voltages. As the high voltages are more sensitive to the effects of radiation, the first block that will be analyzed is the programming and the erasing circuit. As the Flash cell is symmetrical, drain and source terminals can be exchanged. It is then possible to write/erase the cell from one of the terminals (i.e. drain terminal) using PMOS transistors, while read can be carried out from the other terminal (i.e. source terminal) using NMOS transistors.

The programming of the cell is done applying pulses of high voltage to one of the terminals (drain) and to the control gate of the Flash cell, while keeping the other terminal (source) to ground. A certain number of pulses are applied and, then, the Flash cell status is verified by comparing the threshold voltage with a reference cell. In this section, we will only consider the program operation (apply of high voltage) while in the following section, the verification operation will be studied.

Erase of the Flash cells is done applying a positive high voltage to one of the terminals (drain) and a negative high voltage on the control gate. This configuration (Erase 1 in Table 5.1) requires the use of a charge pump to produce not only high positive voltages but also a high negative voltage. This configuration also increases the complexity of the periphery to be designed using the RHBD methodology. It is also possible to redesign the Flash memory in order to use only positive voltages also during erasing (Erase 2 in Table 5.1). If the memory array is inserted in an isolated well, it is possible to erase the Flash cells by applying high voltages to the terminal (drain), leaving the control gate to ground and pulling up to the powering voltage the bulk of the memory cell.

Table 5.1   Voltages applied to the four terminals of the Flash memory cell to program, erase or read it. Vh2 > Vh1 > Vdd

	Vs	Vd	Vg	Vb
Program	0	Vh1	Vh1	0
Erase 1	0	Vh1	−Vdd	0
Erase 2	Vdd	Vh2	0	Vdd
Read	bl/bln	0	Vdd	0

In Table 5.1, the different voltages applied to the terminals of the memory cell to program, erase and read are summarized for Flash cells. Erase 1 considers the use of positive and negative voltages, while Erase 2 considers only positive voltages and the use of the isolating well for the memory array.

If program/erase operations are done from the drain side of the Flash memory, then the source side must be then connected to the ground/Vdd. Considering that the read operation is done from the source side, it is required that the drain side is connected to the ground. To control the polarization of the terminals, it is required to add an additional transistor at each terminal and to connect them to the correct voltage value depending on the operation mode.

It is also important to consider that the voltage of the control gate is different for each operation, which means that word lines decoders should be connected to different polarizations depending on the operation of the circuit. Figure 5.9 represents the final stages for the bit line and word line access of the Flash cell. Bit lines for programming/erase and read/verify are independent, while the word final stage is unique with two possible polarizations.

To date, only the general architecture of the access to the Flash cell and the powering requirements to program/erase and read the memory have been considered. The use of the differential cell structure implies to add

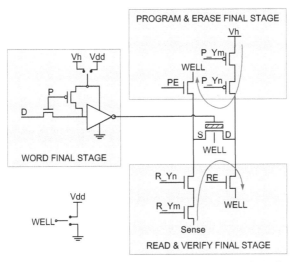

Figure 5.9 Simplified schematic of the program/erase and read/verify final stages connection and word-line final stages. In this example, it is considered the use of a two-step final stage for the word and bit lines. The Well can be connected to Vdd or ground.

some modifications to the periphery. First, we will focus our attention to the programming and then to the erasing circuitry.

The use of a differential Flash cell implies the use of two independent arrays of cells with independent bit-line decoders and a common word decoder to ensure that both cells are selected at the same time. The programming of the differential Flash cell requires the modification of the threshold voltage of only one of the two cells. The selection of the left or right Flash cell depends on the data to be stored, which includes one additional control line to the decoding system. Once selected, the high voltage is delivered only to the required array. Figure 5.10 shows the distribution of the high voltage to the differential flash cell during the programming of a "1" logic bit (left image) or a "0" logic bit (right image).

As already discussed, the use of high voltages should be avoided as much as possible in a radiation environment. This is because the damages to the IC are increased in the presence of higher electric fields. Considering this recommendation, high-voltage (Vh) distribution has been limited to the periphery of the memory array, while the decoding circuitry operates at lower voltages (Vdd), as represented in Figure 5.10.

In the case of the bit line decoding system for program/erase, the switching transistors used are PMOS transistors. To avoid voltage drop in the switching transistor, gates must be driven at the high voltage. The double polarization of the circuit (decoding and switching circuits) requires some adaptive circuitry: a level shifter, which will force the signals from Vdd (decoding) to Vh1/Vh2 (switching) depending on the system status (program/erase). The structure of the level shifter is based on the half-latch structure used in the pads. Figure 5.11 shows the basic structure for the level

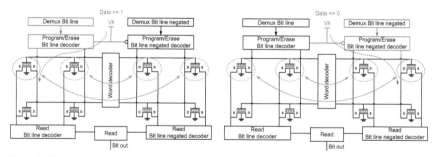

Figure 5.10 Programming high-voltage path to the differential Flash cell depending on the stored data. The left-hand side image corresponds to a logic "1" and right-hand side image correspond to logic "0", following the convention defined in Figure 5.7.

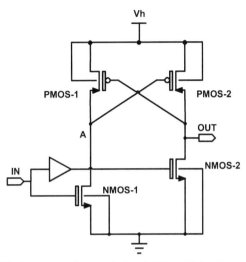

Figure 5.11 Simplified structure of a generic level shifter driving the word lines and bit lines final stages. Input signal is produced by the decoder operating at Vdd, while the level shifter and the final stage can operate at a higher voltage Vh (Vh ≥ Vdd).

shifter, which can be used for the bit-lines. The design of the PMOS/NMOS transistors used in the level shifter are ELT with enhanced guard rings to reduce the effects of the radiation.

The operation of the level shifters is simple: if the IN signal is low, then the NMOS-1 transistor is in cut-off state, while NMOS-2 is in active state. This means that the output node is pulled down to ground, and the PMOS-1 will be turn on. This brings the node A to Vh and cuts-off the PMOS-2 transistor. If the IN signal is high (Vdd), the situation is reversed, and the output node is pulled up to Vh.

So far, the bit-line of the memory has been considered, now the word-line is evaluated. In the case of the word-line, it is possible to use a double stage one operating at low voltage and one at high voltage, with a switching circuit to select between them, or to use a single final stage able to deliver the two polarizations depending on the circuit operation. Considering that all the circuits in the periphery of the array should be designed using ELT with enhanced guard-rings, and the reduced size of the Flash cells, in the case of the word final stage, it is recommended to use a single stage with two possible polarizations, reducing the size of the memory periphery.

The word final stage is represented in Figure 5.12. The basic structure is similar to the one used in SRAM memories. The main difference is that the final stage is connected to a power switch that connects it to the Vdd or

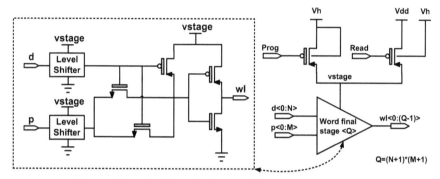

Figure 5.12   Word line final stage with power selection stage.

to the Vh voltage. As in the bit-line stage, the word decoders (which drive the final stage) are powered at low voltage (Vdd), and a level shifter will be used to switch the control signals to Vh if required. In this case, as the same decoder is used for read and write/erase operations, the level shifter has to be modified to keep the same voltage level (Read) or to pull up the signals to high voltage (Prog).

During the write operation, Vh must be applied to the level shifter. In this case, the left PMOS transistor gate signal is low, and the entire supply line of the level shifters is connected to Vh. The right PMOS transistor gate signal is high (Vh), and no current is flowing. During read operation, Vdd should be delivered by the level shifter. In this case, the left PMOS transistor gate signal is high and the right PMOS transistor low. This configuration will supply Vdd to the level shifter. The control signals of the polarization transistors will require the use of an independent level shifter.

When considering these configurations, special care must be taken with the bulk connection of the PMOS transistors involved, and also on the control signals used. For example, the bulk terminal of the PMOS transistors to the low voltage should be connected to the high voltage Vh; otherwise, when the level-shifter-supply node is connected to Vh, the drain-bulk diode would be switched on. On the other hand, during read operation, it is important that the Vh node coming from the charge pump is pulled down to the low voltage (Vdd), but not lower than (Vdd − ΔV), where ΔV is a voltage low enough to avoid that the same diode would be activated.

During the programming of the Flash cells, the neighbouring cells are subjected to a weak erasing condition (Vd = Vh1 and Vg = 0V), producing a decrease of the threshold voltage in the programmed cells. This effect is called erase disturb by programming (EDP) and is a known effect in Flash memory cells. To avoid the EDP, the number of cells sharing the same bit lines is limited to a certain number (maximum number of word lines), which depends on the technology and on the programming/erase conditions.

The erase of the Flash cells is done for an entire bit-line. In this case, all the gates of the bit-line are connected to the ground through the word-line decoder (erase enable control line). A single bit-line is selected, and the drain of the cells connected to the Vh2. The well and the source terminals of all the cells are connected to Vdd. This configuration produces the erase of all the cells in the bit-line. The operation is repeated for each bit-line. A verify operation for each single cell determines if the whole bit-line has been erased correctly.

5.2.3 Read & Verify Circuit Analysis

The difference between the read and verify operations in the differential Flash cell is that during the read operation the two Flash cells of the bit are compared, while during the verify opertion, the cells should be compared with a reference cell separately, as only one of the two cells will have a programmed threshold. This means that two independent circuits will be required. The first part of this section focusses on the read circuit, while the second part focusses on the verify circuit. Both circuits will access the Flash cells through NMOS transistors final stage, which allow high-speed access.

During the read operation, it is important to remember that when a single Flash cell is exposed to radiation, the effective charge of the floating gate will be reduced. This means that if a standard read procedure is done (control gate at Vdd < Vh1), at the beginning of the lifetime of the cell, the programmed cell will not deliver current – except for a certain leakage. While after a certain period of radiation exposure, the cell will have a threshold voltage below the Vdd value and it will be turned on during the read operation.

The use of a differential Flash cell changes the read method, solving the problem of the threshold voltage drift due to TID. In this case, the current delivered with the two cells is compared using a sense amplifier. The reading procedure has a first step corresponding to the equalization (bl and bln are shorted by an NMOS) and pre-charge of the bit-lines to Vdd and then a second step where the cells are selected, and the bit-lines are discharged through the two flash cells. One of the two bit-lines will be discharged in a shorter time due to the cell with a low threshold, while the other will be discharged by leakage current (Vth > VCG = Vdd) or by a small current, lower than the non-programmed cell (Vth < VCG = Vdd). The difference in the current will separate the bit-lines, and the difference is easily sensed with a balanced differential stage with active load. The differential sense stage is recommended to be designed using PMOS transistors since they are less sensitive to the degradation due to TID.

From the reading method previously exposed arises that the output of the sense amplifier will be available only during a certain period of time. This is one of the main differences between this kind of circuits and the SRAMs, which have an internal structure that keeps the data and so the output value fixed. To avoid this problem when using the differential cell, at the output of the sense amplifier, some kind of bistable structure should be added to capture the output and keep it stable for a long period of time.

When considering bistable circuits, many options are available such as flip-flops or latch circuits. The first thing to consider is the fact that bistable circuits are largely affected by transient effect (SET), which can produce upsets (SEU) of the stored bit if the LET is high enough. Special care must be taken to design the bistable to avoid this error during the reading operation of the memory. Typical bistable structures are complex components that require large studies to enhance the performance for radiation environment [8]. A simpler approach to obtain a latching element is to modify a 6T SRAM cell, which has already been designed to be radiation-hard. In this case, the control signal of the pass transistors (Latch) will be used as the control signal to read the sense amplifier output, while the output of the bistable is connected to one of the internal store nodes of the 6T-SRAM cell. In order to be able to load the output line and pad, an inverter or a buffer has to be added to the output to enhance the read time. Figure 5.13 shows a design of the latched 6T-SRAM structure. It is important to note that a Miller

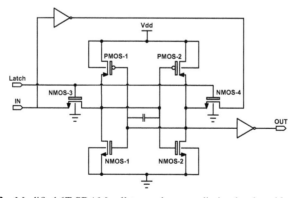

Figure 5.13 Modified 6T-SRAM cell to produce a radiation hardened latch structure.

capacitor will be used to increase the resistance to SET and SEU effects, as already exposed in the SRAM discussion chapter. The use of this capacitor increases the critical charge required (LET) to produce the modification of the information.

The verify mode is part of the programming protocol. To program the differential cell, a pulse of high-voltage signals is applied to one of the two flash cells of the pair to increase the threshold. After a first train of pulses, the verify procedure takes place to check if the cells have been programmed correctly or if they require additional program pulses. The verify procedure is similar to the read operation with the difference that it is conceived to read not the ensemble of the differential Flash cell but the single cells separately. In the read mode, each couple of cells act as a reference for the counterpart, while in verify mode, it is required to have an external reference, such as a band gap or a reference flash cell. The use of a band gap increases the complexity of the circuit and also the sensibility to the radiation effects, since radiation can modify the band gap output during the lifetime of the memory.

The use of reference bit-lines resembles old NVM approach in EPROMs and earlier Flash where it was quite common to use a bit-line embedded in the matrix array to track process variations in the same array. At that time, CMOS technology was not under control as in recent processes and in extended memory arrays, the probability to have a memory cell sinking a different current compared to another cell located far from the first was high. Historically, reference bit-lines were used during read mode and program mode, as well (in particular for EPROMs) and their capability to track process variation and oxides degradation coming from ageing is well known in the literature. Nowadays, the maturity of the fabrication processes has reduced this method to a single reference cell for the whole array.

The reference is used to check if the increase of the threshold voltage is enough or if programming should continue. If the Flash cell has a low threshold, the bit line will be discharged to a voltage value below the reference, while if the Flash cell has a high threshold, after the pre-charge and equalization phases, the bit-line voltage value will be higher than the reference. As in the read operation, the verify operation must be performed at a certain time and the output data must be stored in a latched element to kept data available for the I/O data pad.

As the TID will decrease the threshold voltage of all the Flash cells of the memory, the reference cell should not be programmed during all the lifecycle. For this reason, the reference cell should not be embedded into the array to avoid modifications during the programming or erasing of the memory, thus limiting the threshold drift to the radiation effects.

To ensure that the reference cell presents the same fabrication characteristics as the rest of the array cells, it is required to produce a small array containing the reference. Figure 5.14 shows half of the verify circuit – only one of the two Flash cells is represented. The figure also shows a structure for the reference array. For the differential Flash cell structure, two independent reference arrays and sense amplifiers are required and the output combined to get the result. The array has the same bit-line configuration as the differential Flash cells, contains the same number of cells connected to the bit-line and also contains the same periphery connections of the source and drain terminals. The reference cell is placed in the middle line of the array; few dummy cell lines at each side ensure that during the fabrication process, the variations due to the effect of the different layers will be equal to the cells inside the memory. The vertical position of the reference is also considered. To have a balanced contribution of the reference cell independently of the row that is under verification, the central word-line of the array is used. The rest of the Flash cells of the bit-line will have the gate connected to the ground to reduce their contribution only to the RC load of the source–drain lines, while the reference cell control gate is connected to Vdd. During the verify

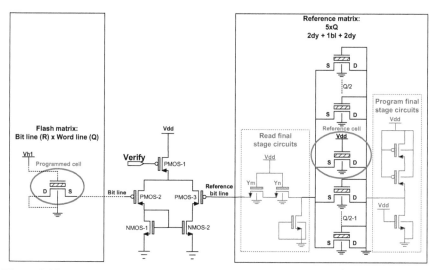

Figure 5.14 Verify circuit and reference matrix cell for one of the two Flash cells of the differential architecture. The dashed lines of the Flash array indicate that part of the circuits has not been designed.

operation, the control gate reference cell is connected to the Vdd, while the control gate of the "programmed" flash cell is connected to the programmed threshold voltage (Vthp > Vdd). The comparison of the current flowing from the two cells will determine if the cell has reached the programmed state.

During the read and verify operations, the timing between the different control signals is a critical factor for achieving a good performance. The distribution of the clock tree represents a criticism in case of radiation environment as each node of the clock tree is sensible to SEE. An alternative system is the design of an asynchronous memory following the structure of the SRAMs. In this case, the control signals timing is produced and controlled using the address transition detection (ATD) circuit and some additional logic electronics. If the address signal or any other control signals (chip enable, write enable, etc.) changes their state, the ATD circuit initiates the control signals needed to perform the required operation (program, erase, read or verify). Following the asynchronous flash structure, a clock signal is necessary only during program and erase to generate the train of high-voltage pulses. It is then possible to confine the clock circuit to the charge pump instead of using a general clock distribution, thus limiting the radiation damages to the program/erase operations, which are rarely performed.

The ATD circuit is activated when an address signals or a control signal (write enable, output enable, chip enable, etc.) are modified. The circuit generates the delayed signals required to pre-charge, equalize, sense the flash cells and latch control required for both read and verify operations. Figure 5.15 shows the general schematic for the read and verify circuit with the ATD generation and the output latch. During the read operation, the two verify sense circuits are inhibited, while during the verify operation, the read sense circuit is inhibited.

Finally, it is important to position the read/verify circuit to have the RC load of the two arrays balanced. This means that the read/verify circuit should be placed in the middle of the two differential arrays of the differential Flash cell.

5.2.4 General Architecture Considerations and Internal Bit Structure

Until now, the particular requirements of the Flash cell have been considered. In this section, the general architecture of the memory considering the RHBD methods already used for SRAMs in previous chapters now applied to the differential Flash memory is discussed. The first step is to avoid the multiple

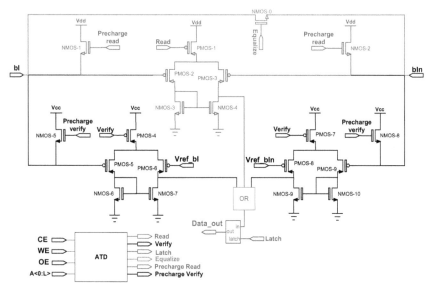

Figure 5.15   Read and verify circuit and the ATD circuit to generate the control signals. In blue is the read circuit, in black is the verify circuit – one for each array of differential cells – in red is the output latch-up and in green is the equalize.

bit upset (MBU). MBU consists of the generation of two or more wrong bits in a single byte due to an SEE. The method used is to divide the different bits of each byte in independent arrays. Each array will have its autonomous decoding and sensing system. This architecture increases the immunity to decoding errors due to SEE and reduces the probability of MBU in a single block of information. An error in a bit can still be produced, but the information can be recovered using an ECC system.

Each bit block contains two arrays of Flash cells with its reference cell arrays. The two arrays produce the specular distributed differential Flash cells. The bit-line decoders are divided into high-voltage components used during program/erase operations and the low-voltage components used during read/verify operations. The division of the decoders suggests the positioning of the two decoders in different positions, one on the top side of the Flash array and the other on the bottom side. This distribution implies the positioning of the bit-line decoders, access bit-line stage and level shifters, if required. In between the two differential arrays, the word decoder together with the power distribution and level shifters is placed. This allows that the decoding signal is distributed to both arrays symmetrically and simultaneously. Also the read and verify circuit is positioned in the middle of the two

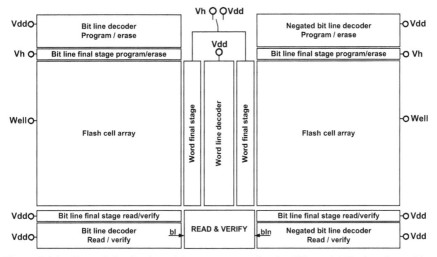

Figure 5.16 General distribution of the components for the differential Flash cell considering the RHBD techniques presented.

arrays. As exposed in the previous section, the read and verify circuit requires the generation of timing-controlled signals that are produced by the ATD circuit. In this case, as the timing control of the read and verify operations is a critical point of the circuit, the ATD circuit is also independent and integrated in each single bit. Figure 5.16 shows a possible distribution of the different components in each bit of the memory.

Until now, a generic Flash cell memory that should present a certain radiation resistance has been considered. To define the real structure of the final memory, some additional considerations have to be considered, such as the size of the memory, the limit on the number of the Flash cells in a bit line, the RC load during read/verify operations and the required timings of the memory. These considerations will be reviewed in the following section, where the methodology is applied to a real device of 1 Mbit.

5.2.5 Case Study: FP7 SkyFlash Project on Rad-hard Non-Volatile Memories

The general specifications for the memory are:

- 1 Mbit NOR Flash memory: $8 \times (2 \times 128$ Kb$)$
- Flash cell characteristics: S-Flash cell ($1.12 \ \mu m \times 1.12 \ \mu m$), EDP limit 128 cells [13, 14].

- Read voltage (Vdd): 3.3 V, asynchronous
- Program voltage (Vh1): 5–5.5 V, synchronous, Number of cycles ~ 100
- Erase voltage (Vh2): 7–8 V, synchronous
- Minimum expected total dose 100 krad (Si)
- SEL immunity ≥ 62 MeV cm^2/mg (Si)
- SEU immunity ≥ 4 MeV cm^2/mg (Si)
- CMOS technology: 180 nm TowerJazz

The first step in the RHBD of the memory is the evaluation of the bit structure. The implementation of the differential cell requires the design of a 2 Mb memory divided into specular distribution of the differential S-Flash cells in each bit, corresponding to two arrays of 128k cells per bit. In the middle of the two arrays, the word decoder and the word final stages are placed. This positioning ensures that the control gate signals arrive at the same time to both S-Flash cells of the pair. The program/erase decoding system and final stages will be placed on the top or bottom side of the array, while on the other extreme of the array, the read/verify decoding system and final stages are placed [12, 15].

Considering the limitation in the number of cells connected in a single bit-line due to the EDP limitation, the arrays should have a dimension of 128×1024 cells. This structure would produce a large array with large RC loads for the word decoders and for the read/verify circuitry. An alternative approach is the division of each bit in four arrays (128×512 cells) instead of two (128×1024 cells). The division in four arrays increases the number of elements devoted to the decoding system of each bit. As the limitation of the number of cells is related to the program operations, the natural option is to duplicate the decoding system of the program/erase high-voltage distribution and keep the read/verify decoding and sense system placed in the middle of the two vertical arrays. According to this architecture, during program/erase operations, the top/bottom arrays are selected separately, while the read mode can access all the word lines at the same time.

On the external side of the S-Flash cells, the reference cell array is placed. The position of the reference array on the external side is done to compensate the RC load of the bit lines of the array. An additional RC load due to the rest of connections present in the word final stage of the arrays is compensated by adding the equivalent RC load to the reference array. The difference of loads is obtained doing an analysis of the parasitic loads at a layout level. Figure 5.17 shows the distribution of the different elements in a single bit.

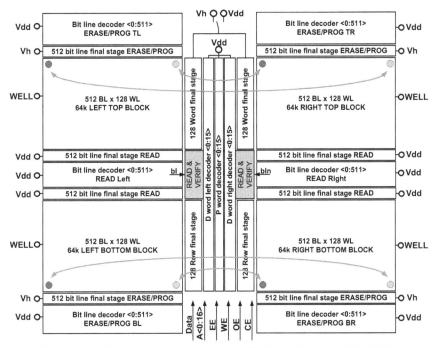

Figure 5.17 Single bit internal architecture of the SkyFlash EU project NVM S-Flash cell.

The partial layout of the 128 Word final stage is shown in Figure 5.18. It can be noted that transistors MP and MR are very huge since they shall connect a supply (either Vh or Vdd) to the final stage and the voltage drop across them must be kept low. These transistors are surrounded by extended guard-ring to limit the SEL risk even considering that Vh generates higher electric fields.

The layout of the single bit block (128 kbit) is shown in Figure 5.19.

Once the bit architecture is defined, the general memory organization can be considered. The first step is the distribution of each single bit following the architecture already defined. The 8 bits can be placed in two columns of four bits and the address distribution lines will be positioned in the middle of the two columns.

This distribution ensures similar delay in the address lines. Also the data lines should have a similar RC load to have an equilibrated component. The bits will be the core of the Flash memory. Around the core, it is distributed the pad ring, containing also the charge pumps that will produce the two required voltages, one for programming (Vh1) and one for erasing (Vh2).

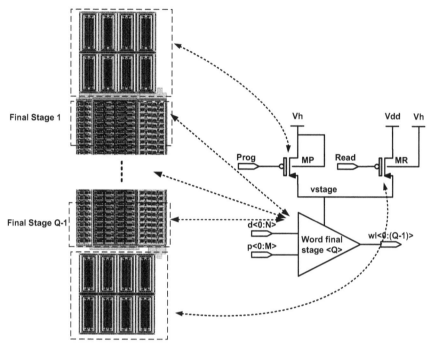

Figure 5.18   Physical layout of the 128 Word final stage (FP7 SkyFlash).

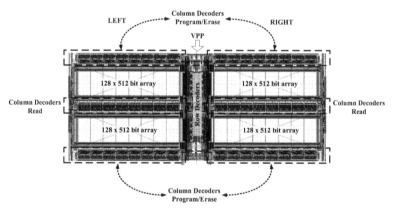

Figure 5.19   Physical layout of the single bit block (FP7 SkyFlash).

Figure 5.20 shows a detail of the "analog" part of the Read & Verify block, where the three sense amplifiers are highlighted at the bottom. Some MIM capacitors have been added to nodes Vref_bl and Vref_bln to compensate the capacitive load given by column decoders final stage (Figure 5.21).

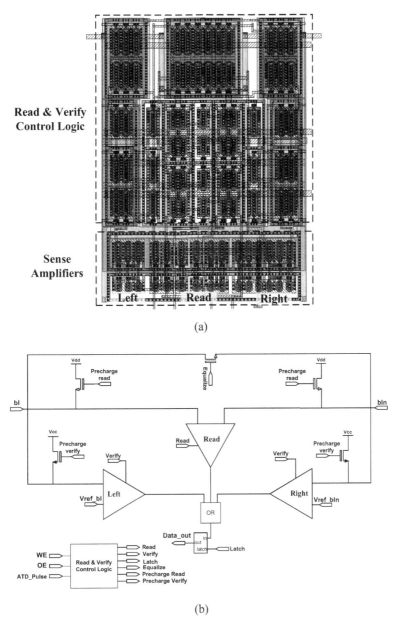

(a)

(b)

Figure 5.20 Physical layout (a) and simplified schematic (b) of the Read & Verify block (FP7 SkyFlash).

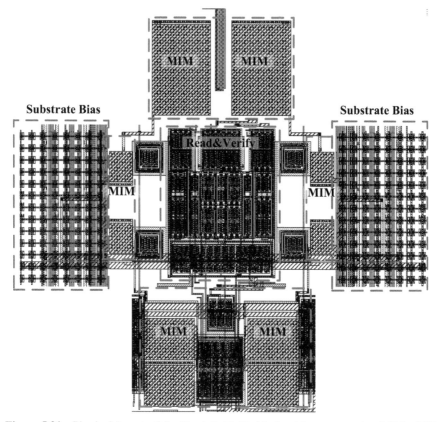

Figure 5.21 Physical layout of the Read & Verify block with compensation MIMs (FP7 SkyFlash).

Substrate bias allows a better polarization of the area around the "analog" part, which is very sensitivity to disturbances.

Figure 5.22 shows the image of the 1Mb S-Flash device designed following the RHBD methodology presented in this chapter. The final size of the device is $3702.57 \, \mu m \times 3188.65 \, \mu m$, corresponding to an area of 11.81 mm². The relation between the array and the periphery is 56% for the array, 44% for the periphery. Of course, this is mainly due to the use of ELT, enhanced guard rings and redundancy required to obtain the radiation hardening.

5.3 Side-Circuits

Two additional components can be considered as side circuits for the Flash memories: the charge pumps and the pad ring. The first element is used only during the programming and erasing operations, where it is required

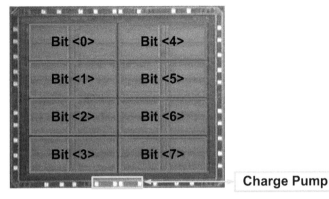

Figure 5.22 1Mb S-Flash memory microphotograph and internal distribution of each bit (FP7 SkyFlash).

to use a higher voltage than the typical polarization. The second component is common to all circuits presenting RHDB specifications; otherwise, the enhancement obtained in the core of the component will be lost because of the effects of radiation in pad region. In the last part of this section, the side circuits used in the real component developed during SkyFlash project are discussed.

5.3.1 Charge Pump Analysis

Many different circuits can be used to design a charge pump reaching the specifications required for the flash memory under consideration. As the study of the charge pumps is not the scope of this chapter, only a general overview of the important aspects involving the design of those circuits will be considered, without going further in detail into the design of a special circuit.

The main problems of the charge pumps under radiation are related to the fact that the circuit is a voltage multiplier, which involves high electric fields. This implies that the probability of charge trapping is higher and that in case of failure the currents involved are also high, which may produce damages in the circuit neighbouring.

The basic components of integrated charge pumps, usually designed with very few components, are PMOS and NMOS transistors, capacitors and diodes. In previous chapters, the radiation effects on some of those components have been exposed, but a short analysis will be presented here for the sake of clarity.

NMOS transistors are sensible to the charge trapping accumulative effect in the gate oxide, which may produce the leakage between the drain and the source in a standard NMOS transistor. To avoid this effect, all the NMOS transistors in the circuit must be designed using edge less transistors (ELT).

PMOS transistors are still sensible to the charge trapping (holes) due to TID, but the threshold shift is in the opposite direction, which allows the use of standard PMOS transistors. The use of standard PMOS transistors allows reducing the area required for the circuit and simplifies the routing of the layout. On the other end, the driving strength in a digital port may be unbalanced if NMOS and PMOS have different shapes. The use of standard or ELT transistors is a designer decision.

All the groups of transistors must be laterally protected using enhanced guard rings to prevent single-event latch-up (SEL) effect. This means that between the n-diffusion and the p-diffusion, it must be added a p-well and an n-well biased at constant voltages.

Capacitors can be produced with multiple types of structures in an integrated circuit, accumulation capacitors, double polysilicon capacitors, MIM capacitors, interdigitated metal capacitors and so on. The possibility of using one or another of them depends on the technology used to build the circuit and, in this case, the compatibility with the fabrication of the Flash memory cell. The main problem on the capacitors is the loss of charge due to the hit of a high-energy particle (SEE) or due to the charge trapping in the isolating media (TID). Those effects will depend on the typology of capacitor selected and on the characteristics of the technology.

A peculiarity of the charge pump design is that it generally uses a multiple stage circuit to obtain the required high voltages. In this case, it is important that the different stages managing different polarizations are protected with additional enhanced guard-rings to reduce the possible inter-device leakage. Therefore, the criteria would be to have independent enhanced guard rings for each group of transistors biased at different voltages.

Finally, the distribution of the clock signals required for charge pumps must be considered. It is possible to integrate the clock generation into the memory ASIC, but it is usually provided externally, as this is part of the full device, and not only of the memory itself. The important consideration is the distribution of the control signals inside the charge pump, which are used in counter-phase (CK and CKN). If these two signals are not synchronized correctly, the charge pump may produce a different output value or a glitch effect. Consequently, special attention must be taken when designing the generation and distribution of the signals following the RHBD rules.

5.3.2 Charge Pump

Following the requirements presented in Section 5.2.5, now it is time to analyze the solution for the charge pump real design. The voltage required for programming is in the range of 5–5.5 V, and the supply voltage is 3.3V. The multiplying factor required is less than 2, which means that a single stage charge pump is sufficient for this project. After an accurate analysis of the different possibilities, the circuit used is a modified CMOS version of the Cockcroft-Walton circuit [16], where the maximum voltage across any device does not exceed the voltage supply. Figure 5.23 shows the schematic of the charge pump, while Figure 5.24 shows the layout. As exposed in Section 5.3.1, PMOS transistors are designed using standard or ELT devices, while NMOS transistors are ELT. Special care has been taken adding the guard-rings for groups of transistors and capacitors. Accumulation capacitors used are 3.3 V gate-to-N-well capacitances divided into an array of smaller capacitances to reduce the N-well resistance effect. Two diodes model the reverse-biased junctions between the deep N-well and the substrate, and between the deep N-well and the P-well are added to model the behaviour of the p-n junctions.

The use of accumulation capacitors has some drawbacks, such as the voltage of the bottom plate limited due to parasitic diodes, and the fact that the reverse-biased p-n junction (well and bulk) is sensitive to the radiation. The first limitation implies that these capacitors allow only single stage charge pumps, while the second limitation must be considered for a radiation-hard design [17].

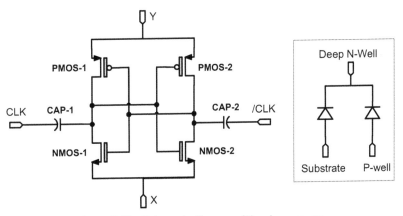

Figure 5.23 Schematic diagram of the charge pump.

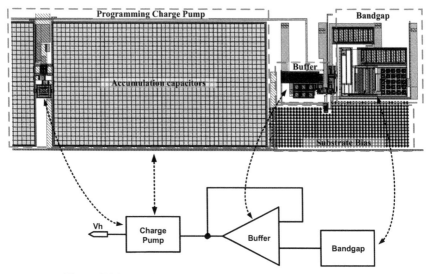

Figure 5.24   Physical layout of the charge pump (FP7 SkyFlash).

5.3.3 Pad-ring Requirements

Although someone could think that the pad-ring is not an important issue in the design of a radiation-hard component, if the pads are not designed correctly, the whole system could fail under radiation event if the core of the ASIC has been designed correctly. The pads can be divided into power pads, I/O pads, input pads, output pads and analog pads. Each of them may contain logic components, diodes and/or passive components. The design of those components should follow the RHBD rules used to design the core of the ASIC:

- NMOS transistors should be ELT.
- PMOS transistors can be shaped either in a regular or an ELT shape.
- Enhanced guard-rings should be used to protect all components from the SEL events.
- Diodes should be designed with the additional proper protections.
- Capacitors/resistors should be analyzed, suitable material chosen and enhanced guard-rings used if required.

Following those basic rules will enhance the radiation hardening of the full component and would guarantee a better performance of the device. Indeed, considering the pad structure, all the pads used in the ASIC of the project were designed using ELT transistors for both PMOS and NMOS transistors.

All the internal circuits where protected using enhanced guard-rings that separate NMOS and PMOS transistors to avoid latch-up events due to SEE.

5.4 Conclusion

In this chapter, we have studied the radiation enhancement of a Flash cell memory considering only the architectural point of view. The introduction of the differential cell to store the information increases the total dose rate that the device allows before failure, reduces the writing frequency and enlarges the lifetime of the component. An accurate analysis of the operation of the device and the circuits used to perform them has bring to the conclusion that separating the high-voltage circuits – more sensitive to the radiation environment – from the low-voltage circuits required during read operation is the best option to protect the system from the radiation effects. The division of the circuits increases the complexity of the circuit but allows us to design the different circuits following the RHBD rules depending on the final application of the component. The guidelines presented in this chapter can be used for any generic Flash memory, considering the special requirements for the Flash cell in use (voltages for programming and erasing, timings, size, EDP, etc.). Special care must be taken also in the design of the side circuits, such as the charge pump and the pad ring, which will affect the final performances of the component. A case study that follows the guidelines presented here, using the S-Flash cell, was developed under the FP7 SkyFlash project.

References

[1] S. Mittal, J. S. Vetter, "A survey of software techniques for using non-volatile memories for storage and main memory systems", IEEE transactions on parallel and distributed systems, Vol. 27, No. 5, pp. 1537–1550, May 2016.

[2] F. Masuoka, M. Asano, H. Iwahashi, T. Komuro, S. Tanaka, "A new flash E^2PROM cell using triple polysilicon technology", International electron devices meeting 1984, pp. 464–467, December 1984.

[3] F. Masuoka, M. Momodomi, Y. Iwata, R. Shirota, "New ultra high density EPROM and Flash EEPROM with NAND structure cell", International electron devices meeting 1987, pp. 552–555, December 1987.

[4] C. Calligaro, A. Manstretta, A. Modelli, G. Torelli: "Technological and design constraints for multilevel flash memories", Third IEEE International Conference on Electronics, Circuits, and Systems (ICECS), Oct. 1996, pp. 1005–1008.

[5] S. Mittal, J. S. Vetter, D. Li, "A survey of architectural approaches for managing embedded DRAM and non-volatile on-chip caches", IEEE Transactions on Parallel and Distributing Systems, pp. 1524–1537, May 2014.

[6] L. Clark, K. Mohr, K. Holbert, X. Yao, J. Knudsen, and H. Shah, "Optimizing Radiation Hard by Design SRAM Cells", IEEE transactions on Nuclear Science, vol. 54, n. 6, December 2007, pp. 2028–2036.

[7] A. Arbat, C. Calligaro, Y. Roizin, and D. Nahmad, "Radiation Hardened 2Mbit SRAM in 180nm CMOS Technology", IEEE-AESS Conference in Europe about Space and Satellite Telecommunications 2012 (ESTEL), Rome, nov. 2012.

[8] A. Stabile, V. Liberali, C. Calligaro, "Design of a Rad-hard library of digital cells for space applications", IEEE International conference on electronics, circuits and systems 2008 (ICECS), pp. 149–152, September 2008.

[9] E. Do, V. Liberali, A. Stabile, C. Calligaro, "Layout-Oriented simulation of non-destructive single event effects in CMOS IC blocks", Proceedings of the 10th European Conference on Radiation and its Effects on Components and Systems (RADECS 2009), Bruges (Belgium), September 2009.

[10] M. Lisiansky and others, "Radiation tolerance of NROM embedded products", IEEE Transactions on nuclear science, Vol. 57, No. 4, pp. 2309–2317, August 2010.

[11] G. Cellere, A. Paccagnella, A. Visconti, M. Bonanomi, A. Candelori, S. Lora, "Effect of different total ionizing dose sources on charge loss from programmed floating gate cells", IEEE Transactions on nuclear science, Vol. 52, no 6, pp 2372–2377, December 2005.

[12] A. Arbat, C. Calligaro, V. Dayan, E. Pikhay, Y. Roizin, "SkyFlash EC project: Architecture for a 1Mbit S-Flash for space applications", IEEE International conference on electronics, circuits, and systems 2012 (ICECS), pp. 612–620, December 2012.

[13] E. Pikhay, Y. Roizin, A. Fenigstein, a. Heiman, E. aloni, and G. Rosenman, "An embedded spacer-trapping memory in the CMOS technology", 2nd ICMTD Proceedings, Giens, France, pp. 195–198, 2007.

[14] Y. Roizin, E. Pikhay, A. Fenigstein, and A. Strum, "Low-cost embedded NVM for power management designs", Ecmag.com – October 01, 2008.

[15] A. Arbat, C. Calligaro, V. Dayan, E. Pikhay, Y. Roizin, "Non-volatile memory for FPGA booting in space", NASA/ESA conference on adaptative hardware and systems 2013 (AHS), June 2013.

[16] J. D. Cockcroft and E. T. Walton, "Production of high velocity positive ions," *Proc. Roy. Soc., A*, vol. 136, pp. 619–630, 1932.

[17] G. Bellotti, V. Liberali, A. Stabile, S. Gregori, "A radiation hardened by design charge pump for Flash memories", RADECS 2013.

6

Radiation Hardness of Foundry NVM Technologies

Evgeny Pikhay[1], Cristiano Calligaro[2] and Yakov Roizin[1]

[1]TowerJazz, Israel
[2]RedCat Devices, Italy

"With a fremen suit in good working order you won't lose more than a thimbleful of moisture a day"

Frank Herbert, Dune

As a specialty foundry, TowerJazz internally developed numerous NVM (Non-Volatile Memory) technologies and evaluated embedded Flash solutions of NVM IP vendors. This included One-Time Programmable (OTP) and Multi-Time Programmable (MTP) logic memories, nitride-based devices and emerging memory solutions, such as MRAM and ReRAM. This chapter summarizes the results obtained during several projects targeting radiation hardening of embedded memories. Special test structures designed to control the radiation influences on the core CMOS and memory devices are discussed. The distinguishing process features of radiation-hard foundry platforms are analyzed and physical mechanisms responsible for radiation hardness of NVM are specified, including the ways of performance improvement. The experience obtained in interaction with customers interested in radiation-hard applications is used for the analyses. In particular, sensors of radiation based on Floating Gate (FG) devices and their radiation immunity are presented in view of possible integration of rad-hard NVM devices and FG structures for radiation measurements at the same chip.

6.1 Introduction

Semiconductor foundries need technology differentiators in CMOS production platforms. One of the main differentiators is embedded non-volatile memories (NVM). Besides the traditional use of NVM for code and data storage, embedded NVM allow flexible design configurations (e.g. trimming of reference voltages, and currents and amplifier offsets), faster time to market and advantages in product testing and yield enhancement. The diverse demands of specialty foundry customers are reflected in memory technologies suitable for different applications, including fabrication of radiation tolerant devices. In many cases, the radiation-hardened memories must also feature low power consumption, enhanced endurance and retention, support memory modules of different sizes with very fast programming, low start-up and read times and allow field programmability and high level of security. While adding the required performance features, embedded NVMs must not lead to changes in the CMOS design kits and IPs [1, 2].

Although semiconductor fabricators started using several types of emerging NVMs (ReRAM, MRAM, ferroelectric, etc.) featuring outstanding radiation immunity, the corresponding technologies are still not mature enough for use in critical applications. Polysilicon floating gate (FG), nitride-based EEPROM and different types of fuse/anti-fuse NVM are dominating in the NVM portfolios of semiconductor fabricators. Storage of data in FG-type NVM is achieved by charge injection into the floating electrode isolated by high-quality dielectrics from the semiconductor substrate and transistor gate [3, 4].

It is well known that radiation environments can have negative effects on CMOS devices, resulting in sudden changes of internal signals, which makes the electronic component to malfunction. Memory modules are especially susceptible to the ionizing radiation. Both the memory cells and the control circuitry can be influenced by different types of radiation. Loss of electrical charge from the floating gate of NVM devices is one of the typical examples. Additionally, prolonged exposure to radiation can change the performance of dielectrics, semiconductors and their interfaces.

The sensitivity to radiation can be tuned by process and design methods, so that FG devices would have enhanced radiation hardness. On the contrary, by increasing the sensitivity of FG devices to radiation, high-sensitivity silicon sensors integrated into the radiation-hard CMOS periphery can be fabricated. This allows realization of a unique combination of analog electronic designs that include sensors and detecting/processing circuitry on the same silicon substrate.

Traditionally, radiation-hard microcircuits were fabricated in specialized foundries owned by such companies as Rockwell, Honeywell and BAE Systems and were intended mostly for defence and aerospace applications. Sensitivity of core CMOS devices to ionizing radiation (total ionization dose – TID) decreases in nanoscaled CMOS due to thinner dielectrics employed. This fact, and also the developed radiation-hard process know-how, allowed several semiconductor foundries, including TowerJazz, to propose technologies for fabricating radiation-hardened devices in their commercial technology platforms. On the top of the process-related immunity to radiation, additional mitigation is possible through Radiation-Hardened-By-Design (RHBD) layout techniques [5]. This allows for faster turnaround time and development of novel chips at substantially reduced costs, which is critical for emerging commodity applications, such as sterilization of medicines, food processing with high doses of gamma and X-ray radiation (to kill pathogens) and electronic devices for radiation medicine, as well as for traditional applications, such as space electronics and devices for nuclear power stations. Non-volatile memories are a very significant element of the radiation tolerant chips and in many cases limit their radiation immunity.

To demonstrate the radiation performance of semiconductor technologies, access to very expensive radiation test facilities (like high-energy heavy-ion accelerators, power gamma sources and nuclear reactors) is needed. The advantage of a foundry is having diverse customers who report verification of the designed products at different facilities. This allows receiving the cross-verification data for similar products irradiated at similar facilities or during different space missions.

The level of radiation immunity currently demonstrated by foundries was previously reported only for digital ICs fabricated in specialized rad-hard facilities. For example, a higher than 15 Mrad Total Ionizing Doses (TID) immunity of products fabricated in one of TowerJazz 0.18 μm technologies confirms the possibility of reaching TIDs of CMOS logics at the level of tens of Mrad and above. Some of TowerJazz customers reported >1 Mrad TIDs also for the analog ICs (e.g. in CMOS image sensors).

In this chapter, we review the NVM used in foundry technologies with the emphasis on the physical phenomena occurring in the corresponding devices under irradiation. We discuss mostly the total dose effects (including TID effects of ion irradiation), since SEE (Single Event Effects) are typically mitigated by the RHBD performed by in customer designs. We start the review from the analyses of degradation effects in CMOS platforms and discuss the process steps that limit radiation hardness. The results are directly applied for the analyses of radiation performance of OTP anti-fuse devices, MTP single Poly FG EEPROMs, SRAMs with integrated beta-voltaic batteries and embedded radiation sensors. This is followed by the description of the radiation performance of nitride NVM (SONOS and NROM – MicroFlash®), including NROM^{TM} low-cost S-Flash and CEONOS versions. This chapter concludes with comments on the radiation immunity of some alternative embedded memory solutions available in TowerJazz fabs, in particular TAS MRAM (Thermally Assisted Magnetic Memory) and TaOx-based ReRAM.

6.2 Physical Phenomena in CMOS Devices Under Irradiation and Their Control

Radiation effects in MOS devices are mostly manifested as degradation of dielectric layers and their interfaces. Charges are captured by the defects initially existing or generated during the irradiation process. Trapping of holes near the Si–SiO$_2$ interfaces and in the dielectric bulk can build surface channels in MOS transistors. This may result in increased leakages and sometimes total failure of electronic circuits. Figure 6.1 illustrates the mechanisms of trapping (a) and source-to-drain leakage path creation at the STI edge of an NMOS transistor (b). Similar leakages can appear between N+ junctions separated by STI.

In the scaled down CMOS technologies, charge trapping (Q_t) at silicon–silicon dioxide interfaces under the gates is usually less pronounced, since $\Delta V_t = Q_t/C$, while C (F/cm^2) is larger for oxides with scaled down thickness. For technologies specially optimized to withstand high TID and other CMOS flavours, the charges accumulated at silicon dioxide interfaces and bulk are also smaller. Figure 6.2 and Table 6.1 present the results of irradiation

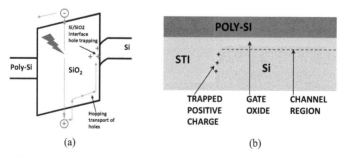

(a)   (b)

Figure 6.1 Mechanisms of trapping (a) and source-to-drain leakage path creation at the STI edge of an NMOS transistor (b).

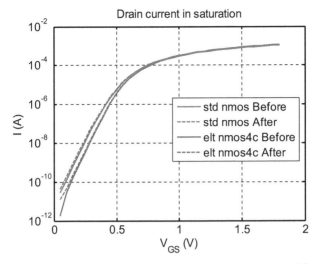

Figure 6.2   Standard and edgeless (donut) transistor fabricated in a rad-hard flavour of 0.18 μm technology before and after 1 Mrad TID.

Table 6.1   Performance of MOS transistors subjected to 1 Mrad TID

	VT (gm max method)	Maximum ID (VDS = 1.8 V)	DVT (gm max method)	Variation in Maximum gm (VDS = 100 mV)	ΔVGS @ ID=1e-8A Sub-threshold (VDS = 1.8 V)
Before	534.7 mV	121.6 mA	−5.4 mV	−3.8%	−11 mV
After	529.3 mV	120.1 mA			

STANDARD MOSFET

	VT (gm max method)	Maximum ID (VDS = 1.8 V)	DVT (gm max method)	Variation in Maximum gm (VDS = 100 mV)	ΔVGS @ ID = 1e-8A in Sub-threshold (VDS = 1.8 V)
Before	534.4 mV	113.01 mA	−1.7 mV	−0.72%	−6.2 mV
After	532.7 mV	113.2 mA			

EDGELESS MOSFET

for transistors fabricated in one of the flavours of 0.18 μm TowerJazz process flow. Results for two types of transistor are reported: (i) Standard MOSFET with the W = 10 μm, and channel length L = 0.18 μm and (ii) Donut (annular, edgeless device, ELT) with no STI periphery; W(perimeter) = 10 μm and L = 0.18 μm.

The devices were irradiated by 10 MeV boron ions and conversion to TID (Gamma) was done according to the published method. The changes of parameters after irradiation are very small, both in saturation and in the sub-threshold operation regimes. The observed differences are not a limitation for digital designs. As mentioned in Introduction, TowerJazz transistor's high radiation immunity was also confirmed statistically by irradiating SRAMs fabricated in the same process flow. The experiments with RedCat Devices 512 kbit SRAMs (RC7C512RHH) were performed at the X-ray facility in Legnaro (Padova), Italy, in passive and active modes. No errors were observed for absorbed doses of above 15 Mrad (Si). Radiation immunity of the demonstrated level was previously reported only for digital ICs fabricated in dedicated foundries specialized rad-hard technologies. It is necessary to mention that not only 1.8 V transistors but also input–output edgeless 3.3 V devices withstood the mentioned doses.

Analog devices fabricated on the same platform remained functional at radiation doses >100 krad. This was confirmed by several image sensor customers targeting rad-hard applications Also, 4-Bit ADC have been designed in the mentioned 0.18 μm technology. The prototypes have been tested under Co-60 and showed radiation immunities >300 krad (Si).

Two types of optimizations were performed to achieve this level of radiation immunity:

- STI optimization to increase the filling dielectric material density and decrease the amount of moisture;
- special capping layers [6] in the back-end dielectric stacks introduced to decrease the amount of hydrogen ions generated.

The importance of hydrogen for the performance of CMOS devices is well known in the semiconductor industry. A certain amount of hydrogen at the Silicon–Gate Oxide interface is needed to passivate the surface states (by forming Si–H complexes with dangling bonds). At the same time, excess amounts of hydrogen are responsible for device instabilities. For example, excess hydrogen may result in floating gate memory retention loss [7]. In particular, higher radiation hardness of BPSG with lower amounts of hydrogen was associated with lower amount of hole traps related to hydrogen complexes. At the same time, hydrogen atoms are also known to interact with the bulk silicon defects created by ionizing radiation [8]. Passivation of radiation defects by radiation-activated hydrogen originating from inactive hydrogen containing species was reported. To ensure high radiation immunity, optimization of the hydrogen content is required. In the rad-hard process flow, control over the amount of generated hydrogen was done that allowed a significant decrease or eliminating hydrogen production during the formation of metallization layers of the back-end metallization structure ([6] and Figure 6.3).

The amount of hydrogen atoms/ions produced during back-end processing depends on the metallization scheme. When forming an inter-level dielectric (ILD) layer, TEOS-based (tetraethylorthosilicate) oxides (e.g. Undoped Silicon Glass, USG or Fluorinated Silicon Glass, FSG) capped with low-moisture-content oxide layer was used such that the cap layer served as an etch-stop for subsequent aluminium over-etch (after aluminium etch was completed, the low-moisture-content oxide cap layer still had a sufficient thickness). The use of TEOS-based oxide to form the ILD layer assured superior coverage of underlying metallization structures due to its excellent

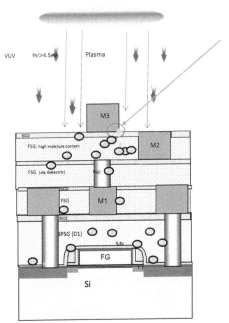

Reaction of Al with H_2O during the overetch: both atomic hydrogen (H) and charged ions H+ are generated

Reaction is used for passivation of Si-SiO_2 interface. During final anneal in Hydrogen (forming gas : N_2/H_2), Hydrogen atoms move to Si-SiO_2 interface and passivate surface states generated as the result of plasma damage.

Figure 6.3   Reduction of hydrogen atoms in a rad-hard process flow.

gap filling characteristics. By forming cap layers using, e.g. silane oxide with low moisture content, control over the amount of hydrogen generated during back-end processing was provided.

6.3 Radiation Hardness of CMOS Logic Memories

Several types of single poly low-cost NVM solutions are used in foundries. Besides the radiation-induced hole trapping, defects in the gate silicon dioxide may result in the formation of SILC (stress-induced leakage currents) percolation paths responsible for the retention performance [7]. In the case of floating gate memories, gate oxides (GOX) having thicknesses above 70 Å (typically 85110 Å) are used to exclude effects connected with SILC.

Decreasing the thickness of the single poly FG memory oxide increases the probability of the leakage path creation. Below we discuss the radiation hardness of single memories from both the view of written data loss (FG discharge under radiation) and CMOS materials degradation.

6.3.1 Single-Poly EEPROMs

Radiation hardness of single-Poly NVM is presented in an example of Y-Flash developed in TowerJazz for power management and special applications requiring small and medium-sized memory arrays. The Y-Flash memory cell consists of an NMOS transistor with a floating gate capacitively coupled to source and drain, and the floating gate extension overlapping the drain area outside the channel region, so that the capacitance of the FG to drain is higher than the capacitance of FG to source (Figure 6.4). In programming, the voltage is applied to drain while the source is grounded. The drain voltage is transferred to the FG, thus facilitating programming by channel hot electrons (CHE). In erasing, the drain is connected to zero potential while positive voltage is applied to the source. This results in the injection of band-to-band tunnelling (BBT) holes into the FG.

The channel area of the device is implanted to increase the lateral electrical field near the drain using implants through the masks existing in the core CMOS process flow. The readout transistor is a standard CMOS device. Both transistors have a common floating gate. No special control gate such as that used in typical EEPROM devices is needed. The operating regimes of the Y-flash cell are presented in Table 6.2.

Device programming is self-convergent, which ensures tight Vt distributions in the memory arrays. No special high-voltage (HV) devices are used in

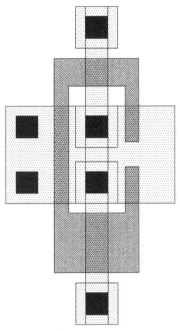

Figure 6.4 The standard cell ($\sim$3 μm^2). Injection transistor channel length, Li = 0.3 μm. Read-out transistor channel length Lr = 0.6 μm. The width of the extension area (LV NLDD, Low-Voltage N-Halo Light Doped Drain) Lf = 0.98 μm. Extension finger width, Wf = 0.14 μm.

Table 6.2 Y-Flash cell operation conditions

	Timing	Voltages	Currents	
READ		2.5 V +/–5%	$\sim$ 20 uA	
PRG	10 ms	6 V +/–5%	250 uA	
ERS	20 ms	8 V +/–0.25%	.300 nA	
		Terminal (2C)		
Operation	Drain	Source_INJ	Source_READ	Substrate
Read	2 V +/–5%	floating	0	0
PRG	6 V +/–5%	0	floating	0
ERS	floating	8 V+/–0.25 V	floating	0

the array designs. Voltages higher than Vdd are supplied using a cascode connection of standard MOS transistors. This significantly increasing the radiation hardness of the corresponding modules. Drain coupling ratio is a principal parameter for Y-Flash cell operation since the device does not have a special control gate. The programming and readout voltages are applied to the interconnected drain terminals of injection and read-out transistors coupled to the same floating gate. The memory cell area in Figure 6.4 is about 3 μm^2 in 0.18 μm design technology.

From the point of view of information storage after irradiation, single-poly EEPROM devices have strongly increased radiation immunity compared with their double-poly counterparts. This is mainly connected with the absence of thick inter-poly dielectrics. In the design, special care also must be taken to exclude the influence of STI/field oxide regions. These regions comprise thick oxide where numerous electron–hole pairs discharging the FG are generated by radiation. Although the contact of FG with STI is limited to a relatively small area (<0.5 μm^2), radiation hardness of this device flavour (60% of information charge) is at the level of 100 krad only. Special radiation-hard Y-Flash cells were developed where contact of the FG with STI was excluded. Figures 6.5(a) and (b) show the layout of these devices and their reaction to irradiation by ^{60}Co gamma rays. The studied cells had varying coupling ratio of drain over the FG defined by the different number of extension rings (squares). The discharged state of cells was defined after 30 min UV erase using a standard EEPROM eraser. All cells were initially programmed to Vt ~4.2 V. The cells with low coupling ratio (one extension ring) had initial higher Vt (lower coupling ratio of drain over the FG). Figure 6.5(b) shows three graphs each for different absorbed doses. The results coincide for three cells irradiated to the same dose. It is seen that radiation hardness (defined dose at 60% window closure [8]) corresponds to the total absorbed doses (Si) above 1 Mrad. This is much higher than typical ~100 krad doses withstood by standard EEPROMs. By using differential cells, the maximum dose when NVM remains programmed can be increased to several Mrads (Si).

Compared with the device shown in Figure 6.4, the radiation-hard Y-Flash has larger cell area (5–15 μm^2/bit). Nevertheless, it is perfectly suited for small/medium-size NVM arrays. It is necessary to note once again that larger arrays are also limited by the appearance of SILC leakage channels.

6.3.2 GOX Anti-Fuses

Another type of popular single-Poly memory in rad-hard applications is gate oxide anti-fuses [9]. These devices, especially one-time programmable (OTP) NVM, are widely used in various applications such as field-programmable gate arrays (FPGA), static RAM (SRAM), small memory blocks for trimming and Chip-ID and even for code storage. The GOX anti-fuse NVM technology

(a)

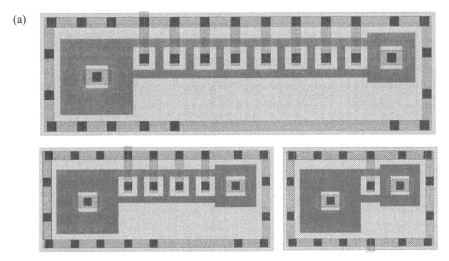

(b)

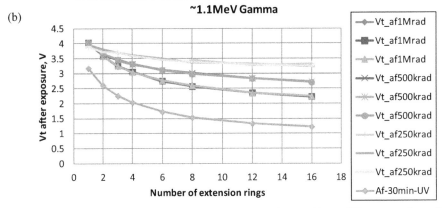

Figure 6.5 (a) Y-cell field-less cells with different drain couplings (different number of extension rings); (b) Response to gamma radiation of rad-hard Y-Flash cells.

employs gate oxide breakdown achieved by applying a high voltage (at the level of 9–10 V) to the gate of a transistor with respect to the source or silicon bulk. Anti-fuse technology allows small bit cell sizes with typically only 2 MOS transistors needed per bit. Programming can be done at sort operation or in the field (if high-voltage charge pumps are available on the chip). The GOX anti-fuse memory was broadly discussed for radiation-hard applications, especially by FPGA designers [10]. It was assumed that in case of programming success, no changes in programmed bits would appear up to very high doses of radiation. Unfortunately, this is not always the case. Even if SILC channels in the non-programmed bits would not appear under radiation (low-voltage thin gate oxide, e.g. ~30 Å in 0.18 μm technology is typically used in anti-fuse modules), there are other reliability limitations in GOX anti-fuse NVM. The yield of programming (breakdowns) is usually below 100% for high-quality gate oxides used in radiation-hard technologies [11, 12]. Existence of poor "breakdown" (quantum point contacts, SILC channels, etc.) can occur if the stress current is not sufficient during the breakdown [13]. The gate oxide may become non-conductive again after bakes. This is usually associated with the defect curing because of the rearrangement of conductive filaments. The result of radiation influences on such breakdown channels is not clear. Also, as mentioned above, high-voltage transistors are needed in gate-oxide anti-fuse memory modules. The high-voltage charge pumps, access and blocking transistors are more vulnerable to radiation damage because of the thick GOX employed in corresponding devices. Probably, this was the reason that companies that traditionally used anti-fuse memories in rad-hard designs have been now investigating new opportunities [14]. In any case, special design features (ECC, redundancy) and qualification that employs large amounts of anti-fuse memory modules is necessary to justify their use in specialty applications requiring radiation hardness, similar to the FG memories.

Bringing the single-poly NVM immunity to radiation to the level of TID of the core CMOS process (>15 Mrad) is possible within an approach described in [15], which suggests a combination of an SRAM and a beta-voltaic source.

6.3.3 Radioisotope-Powered Memory

A radiation-hard non-volatile memory can be made using a battery-backed SRAM, if high radiation hardness of SRAM was already achieved. Unfortunately, this approach is not suitable for most of applications since the use of the external battery diminishes the advantages (similar to radiation-tolerant solutions where protective shields absorbing radiation are used to achieve high TID).

The approach proposed to foundry customers suggests the beta-voltaic source not as the main power supply of SRAM [16], or other low-power circuit with memory features, but only for compensation of the leakage in the passive mode and only in the memory array itself. In the readout mode, SRAM is run from an external ("normal") power supply. Only in the stand-by mode, an option of the beta-voltaic power (tritium-based) is added. SRAM

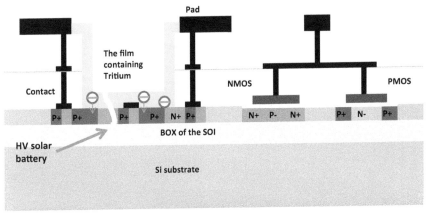

Figure 6.6   Integration of beta-voltaic source with HV photodiodes.

(or other device) can be run at voltages much lower than Vdd (close to Vt of CMOS transistors), which requires currents at the level of nano-Amps only. This makes integration of beta-voltaic sources with much lower power (of the order of single nano-watts), containing very small amounts of tritium, practical. High-voltage photovoltaic cells available in the core CMOS process at no additional cost [17,18] are used in beta-voltaic cell operation, except that instead of converting light to make a current out of a solar cell, beta radiation (electrons) from tritium is used. The beta-voltaic material is "buried" inside the chip by making special cavities after the main CMOS processing. This allows also to exclude the parasitic radiation effects in the sensitive areas of the chip (beta-electrons are absorbed in the beta-voltaic battery area only (Figure 6.6)

The beta-voltaic battery is "spread" over the chip, meaning that numerous beta-voltaic sources are formed inside the chip and internally connected at the level of single SRAM cells, or groups of SRAM cells, as shown in Figure 6.7.

6.3.4 Silicon Nitride-Based Memories

The above-reported record radiation immunity of single-poly memories is limited to small to medium-sized arrays. As in the case of all non-volatile memories based on the polysilicon floating gate principle, radiation-induced defects in the gate silicon dioxide may result in the formation of percolation paths responsible for the gate leakage current [19]. Even one high-resistance leakage path strongly endangers the retention capability

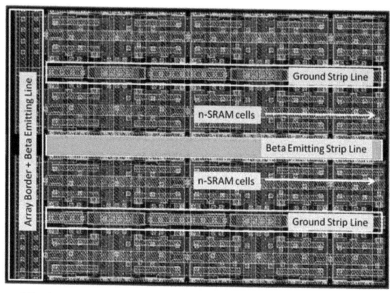

Figure 6.7 Array arrangement of an SRAM device with tritium-embedded distributed sources.

of the Flash memory with a conductive FG. The effect depends on the FG memory bottom oxide, since the probability of the leakage path creation strongly increases with decreasing oxide thickness (TOX). Most FG memories have 80–85 Å minimum bottom oxide thickness to exclude electron tunnelling from the FG through single traps created during cycling and irradiations. The performed characterization shows that SILC-related effects can be registered already for 64 kbit Y-Flash arrays with 70 Å GOX. In case of battery-backed NVM arrays, there are limitations connected with power limitations of practical (small-area) beta-voltaic sources. Thus, the above-described beta-voltaic solution is practical, e.g. for FPGAs, but not for high-volume memories.

Nitride-based memories are free from the above-reported limitations of FG EPROMs. They were reported to have better radiation immunity compared with PolyFG devices due to storage of information charge at traps in the nitride layer [20]. Even if a leakage channel is created, this results only in a small decrease of the charge in the silicon nitride layer. Silicon–nitride–oxide–silicon (SNOS) and silicon–oxide–nitride–oxide–silicon (SONOS) structures have been widely used in the non-volatile memories (NVM) intended for operation in the radiation hardened military and space systems [21]. Some of the SNOS transistors remained functional (though erased) after irradiation to a total ionization dose of more than 1000 Mrad. Nevertheless, to reach this level of radiation immunity at the NVM module level, special design should be done also in the CMOS periphery, especially in the circuits that supply high voltages for programming and thus feature thick gate oxides.

The dominant SONOS-like technology at the NVM market is NROM (nitride read-only memory), fabricated in TowerJazz under the name microFlash®. It features significant memory density increase compared with the classical SONOS by storing information as charge packages in silicon nitride at both transistor channel edges. Programming of NROM (by channel hot electrons – CHE) and the read operation are performed in opposite directions. The memory array has a field-less cross-wise organization. Unlike other nitride NVMs, the bottom oxide of NROM (BOX) is relatively thick ($\sim$70 Å), thus ensuring excellent retention, suppressed read disturb and allowing much easier scaling compared with the PolyFG memories.

NROM technology flavours that guarantee optimal radiation immunity of NVM products were selected after analyzing the dominant physical mechanisms responsible for the functionality and reliability of NROM products in radiation environments. Below we discuss the changes happening under irradiation in the NROM memory arrays and report the correlation of the array performance with the characteristics of irradiated single cells.

Different sources of radiation were used in the laboratory modelling of radiation influences. Most of the experiments were performed by irradiating the NROM-embedded modules with γ-rays from ^{60}Co sources and by 10 MeV 11B ions (LET = 3.9 MeV cm^2/mg) in the range of absorbed doses 10 krad (Si) 1 Mrad. Functionality of NROM was controlled during and after irradiation. Single cells and microFLASH@ 4 Mbit embedded memory modules (Topaz) fabricated in 0.18 μm technology were tested. The total volume of the 4 Mbit embedded Flash memory is divided into 16 equal sectors. A sector includes a contactless array with the cross-wise word and bit lines (256 × 1032, respectively) and CMOS periphery. The last contains bit-line (BL) and word-line (WL) drivers, power switches and charge pumps, sense amplifiers and reference cells. Each sector is controlled separately. It can be programmed/erased or readout independently of other sectors. Such internal structure of the product significantly widens its testability in radiation environments.

A single cell with typical dimensions W × L = 0.18 × 0.42 μm^2 is the basic element of the full-sized product. The dielectric properties of the ONO (oxide–nitride–oxide) stack prevent charge loss from silicon nitride and assure bit separation and high endurance/retention of the memory cells. Unlike the traditional SONOS with Fowler-Nordheim injection, program/erase operations in NROM are performed using CHE for electrons and band-to-band tunnelling (BBT) for holes. To emulate the features of a real NROM cell in a product, each single cell in the test chips was placed in the middle of a 3×3 mini array. To control the possible effects of lateral charge spread in the nitride layer, single bit programmed pattern (AAAA-hexadecimal code) was taken as a base (4 Mbit Topaz NVM modules). Devices were programmed and measured before and after each irradiation. Array threshold voltage distribution (Vt is the gate voltage at which the cell current equals the current of the reference cell) were monitored. In single-cell measurements, the threshold voltage shift (ΔVt) and leakage currents were measured in forward and reverse directions. Irradiated ONO capacitors were studied in parallel with NROM cells and products to control degradation effects at the interfaces (by C–V and charge pumping techniques). TID tests of the embedded modules in the active mode were performed using a special auto-diagnostic board designed using radiation-hard electronic parts. The board operated in the remote-control regime could be placed in the chamber where irradiation was carried out.

The results of the experiments have shown that NROM radiation tolerance was to a great extent controlled by the performance of dielectrics filling the gaps between the WLs in the cross-wise memory arrays (spacer dielectrics and a part of the pre-metal dielectric). The reported results are optimized for the high radiation immunity NROM memories. The programmed embedded modules withstood above 100 krad TID irradiations. For above 1 Mrad, the information could be still read out, but failures in the high-voltage circuitry (170 Å GOX) in the CMOS periphery were observed. This periphery used standard (not annular) transistors and was not specially designed to support high radiation immunity.

Figure 6.8 shows the results of 80 krad ^{60}Co irradiation of the Topaz-embedded NVM module. The total program and erase margin loss is $\sim$1 V. The margin loss does not depend on the regime of the readout. It means that most of the programmed cells lose their charge (or positive charge is trapped), but the mechanism of radiation-induced charge loss is not sensitive to the bias applied to the cell terminals.

The array is programmed in an AAAA (CKB-checkerboard) pattern. In the CKB pattern, the number of cells in programmed and erased states is equal. Under irradiation, the distribution curves of both programmed and erased devices shift towards each other, thus decreasing the window margin

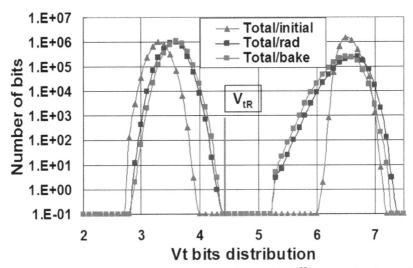

Figure 6.8 Topaz array: programmed, after irradiation (80 krad, ^{60}Co source) and annealing at 150°C for 24 h.

in the NVM array. The post-irradiation retention bake additionally decreases the program margin ($\sim$0.1 V) and erase margin ($\sim$0.05 V). The observed shifts correspond to charges at the periphery (edges) of the memory cells. No degradation of memory retention was observed after irradiation. The NROM memory remains fully functional: no readout mistakes were observed, and the array could be reprogrammed in the standard regime.

The analysis of cells that form the tail of Vt distribution (Figure 6.8) shows that these cells are scattered uniformly over all sectors of the memory product and their location does not depend on the position of a cell in the memory array (centre, or periphery). The programmed (by F–N injection) SONOS capacitors irradiated together with the NVM modules showed only $\sim$0.5 V V_t loss after absorbed doses of up to 5 Mrad (Si). The measurements of single cells irradiated with 5 Mrad indicated that most of them remained functional to the observed tail related to charge trapping in the oxide filling the spaces between the word lines in the NROM arrays. Measurements of single cells revealed increased leakage currents in cells that had decreased Vt (Figure 6.9) for different types of radiation.

An additional manifestation of this effect was the increase of the offset (drain leakage) current in the programmed/erased memory cells. Both drain and source currents were measured. They were equal, thus showing that the leakage was between source and drain (Figure 6.10)

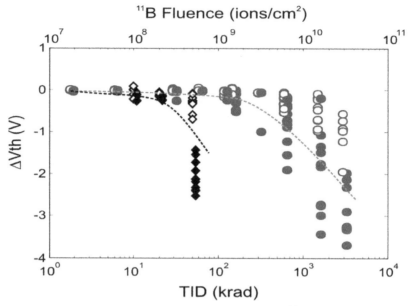

Figure 6.9   Vt decrease as a function of the TID (in krad) and [11]B fluence for programmed cells: γ-rays (black ♦) and [11]B ions (blue •) for forward (open symbols) and reverse (closed symbols) readout conditions.

The leakage is partially annealed during the post radiation bake at 150–250°C. The effect is similar to generation of positive charge in the low-quality isolation silicon oxides (field oxide – FOX or shallow trench isolation – STI), known as one of the main degradation mechanisms in the irradiated CMOS circuits with standard MOS transistor design.

The above results show that high-volume radiation-tolerant embedded modules are feasible using NROM technology. If a level of several Mrad is required, the array and CMOS periphery must be redesigned. MOS transistors having high radiation immunity compared with standard devices with STI edges must be used, e.g. MOS transistors having a fieldless (donut or ELT) geometry. This was done by TowerJazz and RedCat Devices to fabricate OTP high-volume memory modules allowing to withstand TID at the level of 1 Mrad and above). Radiation-hard OTPs are used for booting software and star maps storage in space missions and play a crucial role in the entire life cycle of a satellite since they contain fundamental data used to a safe restart (in case of switch off due solar activities) or for a correct positioning

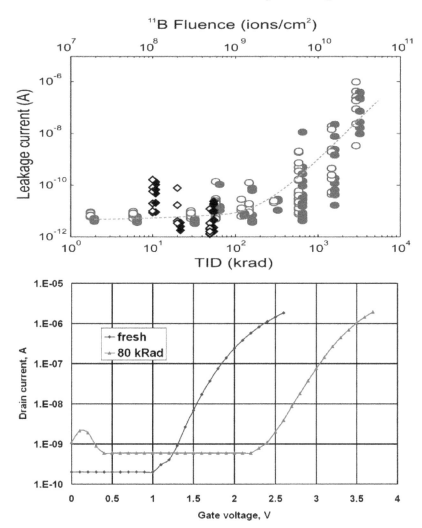

Figure 6.10 Leakage currents in the memory cell after irradiation: (a) different absorbed doses; (b) leakage of programmed cell.

along the orbit (star map for positioning). As mentioned above, old anti-fuse technologies allow only very low storage capacity (max 256 kbit). The complexity of functions required from modern satellites requires an increased capacity of operating systems used to boot the satellite. Moreover, anti-fuse technology has the above-mentioned (Section 6.3.2) limitations. OTP must be programmed only one time (before the launch of the mission) and then it must guarantee resistance against radiation only during reading.

Several low-cost alternatives to microFlash were developed to broaden the application range of the embedded modules. One of the solutions is S-Flash (Figure 6.11).

It employs charge trapping in the silicon nitride spacers of an MOS transistor with skipped LDD (Light Doping Dose) implant and thinned pad oxide layer under the spacer Silicon Nitride. The S-Flash cell can be programmed and erased by voltages below 5 V; thus, no special high voltage devices or circuitry are required (only radiation-hardened 3.3 V transistors of 0.18 μm foundry process flow). Programming is performed by channel hot electrons or secondary electrons, while erase is done by band-to-band tunnelling holes accelerated in the drain junction similar to the way this is done in NROM. The spacer design and junction implantation profile were optimized to achieve the

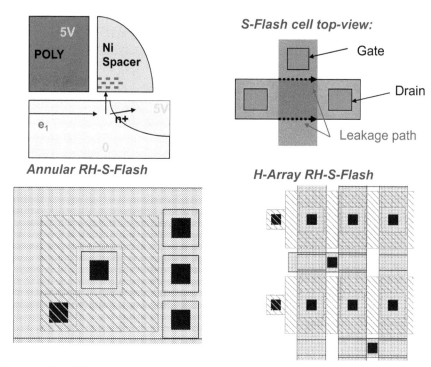

Figure 6.11 S-Flash memory: (a) principle of S-Flash, (b) standard layout of S-Flash memory cell and (c) donut S-Flash and corresponding array.

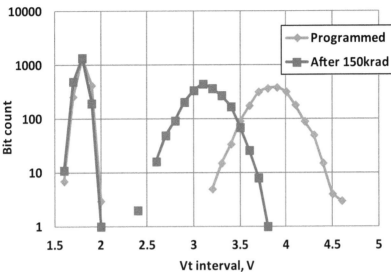

Figure 6.12 Radiation immunity of S-Flash.

highest efficiency of hot carrier generation (high lateral fields and optimized position of hot carrier generation region). This optimization is performed by adding only one additional mask to the standard CMOS process flow. Radiation tolerance of standard S-Flash (4k array) is presented in Figure 6.12. It withstands >100 krad TID. Like the case of NROM, Vt decrease after irradiation relates to positive charge trapping in silicon dioxide isolating the memory cells in the array.

6.4 On the Chip Tools for TID Radiation Effects Control

Often, the radiation-hardened foundry technologies are modified to include demands of diverse customers to functionality, integration of novel materials, additional requests for operation in harsh environments, etc. Thus, there is a need of advanced tools for characterization of the process technologies that are selected for fabrication of radiation-tolerant chips. In this section, the semiconductor devices for fast and direct monitoring of TID implemented in TowerJazz are described. Operation principles of these devices utilize the fact that ionizing radiation produces energetic charges in semiconductors. These charges can lead to accumulation of static charges in dielectrics and/or to discharge the floating gate of NVM devices. To evaluate charge trapping in STI and Si nitride spacers, various MOS transistors were built in the standard CMOS technology, as described in Section 6.4.1. To evaluate the TID in broad ranges of absorbed doses, various FG devices were utilized, as described in Section 6.4.2.

6.4.1 MOS Structures for Monitoring of Radiation-Induced Charges in Dielectrics

A known method to investigate the electric charge trapped in an insulator is to use an MOS structure with this insulator as the gate dielectric. To study charge trapping in STI and Si nitride spacers, various MOS transistors were built in the standard CMOS technology, as described below.

The first two types of structures are NMOS transistors with Polysilicon gate and N-well drain and source (Figure 6.13): one with P-well substrate and another "native" (non-implanted starting material) p-type, 1–2 Ohm*cm substrate.

The second two types of structures are PMOS transistors with polysilicon gate and P-well drain and source: one with N-well substrate and another with the native substrate (Figures 6.14(a) and (b)). The N-well substrate is implemented using N-well and deep N-well implants of standard CMOS process. Id–Vg curves show threshold voltages of about −40 V and −5 V for N-well and native transistors, correspondingly (Figure 6.14(c)).

The third two types of test-structures are "inversed" STI (I-STI) NMOS transistors, in which the polysilicon plays the role of a "substrate", and P-well of standard CMOS plays the role of "Gate". The drain and source are represented by N+ implants in Polysilicon (Figures 6.15(a) and (b)). The "P-well" is implemented by doping the channel region of Poly by P-LDD implant ("lightly doped drain" extensions of standard CMOS technology). The "native" device is implemented by leaving the channel region of Poly undoped.

The mobility of charge carriers in polysilicon is many orders of magnitude lower than those in crystalline silicon. Therefore, the drain current of "inversed" STI transistors (Figure 6.13(c)) is much lower than in the case of "conventional" STI transistors. In addition, the difference between "native" and "P-well" I-STI transistors is larger than in the "conventional" STI case, because the dose of P-LDD implant utilized to form the "P-well" of the I-STI NMOS is ∼20 times higher than the dose of the standard P-well.

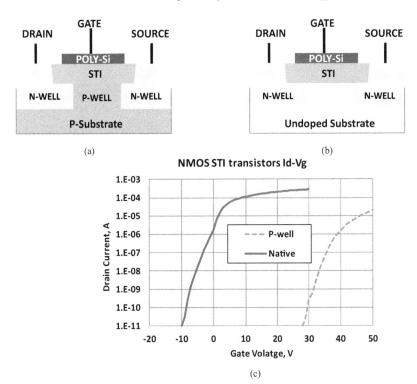

Figure 6.13 NMOS transistors with STI gate dielectric: (a) with P-doped substrate; (b) with undoped substrate; (c) typical Id-Vg curves. The lateral dimensions of NMOS devices and gate dielectric thickness are the same as for C-Sensor control capacitor. Poly-Si doping dose is ~1e 15 cm^{-2}, N-type.

The fourth type of the test-structures is I-STI PMOS transistors, implemented by a similar method to I-STI NMOS transistors mentioned above. The drain and source are formed by P+ implants in polysilicon (Figures 6.16(a) and (b)). The "N-well" is implemented by doping the channel region of Poly by N-LDD implant ("lightly doped drain" extensions of standard CMOS technology). "Native" devices are implemented by leaving the channel region of Poly undoped. Typical Id–Vg curves of I-STI PMOS transistors are shown in Figure 6.16(c).

The fifth group of devices is Poly-to-Poly transistors consisting of two polysilicon bars, one of them heavily doped and playing the role of a gate, while the other plays the role of substrate (channel at the sidewall of the Poly bar). The doped regions at the edges act as drain and source terminals (Figures 6.17(a) and (b)).

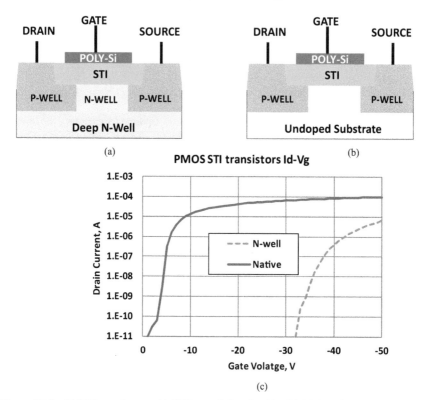

Figure 6.14   PMOS transistors with STI gate dielectric: (a) with N-doped substrate; (b) with undoped substrate; (c) typical Id–Vg curves. The lateral dimensions of PMOS devices and gate dielectric thickness are the same as for C-Sensor control capacitor. Poly-Si doping dose is $\sim$4E 15 cm^{-2}, P-type.

There are two Poly-to-Poly structures, one with N+ gate and source/drain diffusions and the other with P+ gate and source/drain diffusions.

To study the buildup of charge in dielectrics surrounding the FG, the above-described test structures were exposed to different doses of gamma radiation (^{60}Co source; dose rate, 0.85 Gy/h; radiation facility of the Ben-Gurion University of the Negev, Israel) in a passive mode (unbiased). Id–Vg curves were measured before and after the exposure. To account for the spread of the measured Vt change (before and after irradiation) for each test case, we use the average values of Vt change. The analysis of the obtained results is summarized in Figure 6.18.

In Figure 6.18, we see that NMOS STI transistor with P-well substrate shows significantly higher Vt change than the rest of the devices, in particular

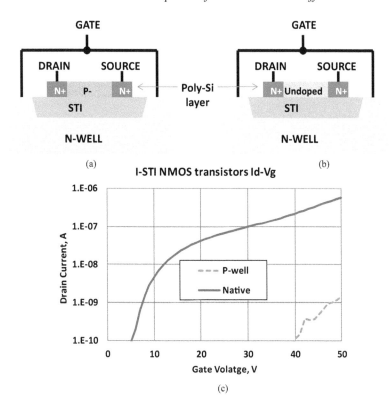

Figure 6.15 "Inversed" NMOS transistors with STI gate dielectric: (a) with P-doped "substrate"; (b) with undoped "substrate"; (c) typical Id–Vg curves. The lateral dimensions of I-STI NMOS devices and gate dielectric thickness are the same as those for C-Sensor control capacitor. Poly-Si doping dose is $\sim 1E$ 14 cm^{-2}, P-type.

vs identical transistors with undoped substrate. The difference is attributed to P-well doping.

A buildup of positive charge at Si/SiO_2 interface of MOS structures due to exposure to ionizing radiation observed in this work agrees with numerous published results. Of all studied test structures, NMOS STI transistors had poor radiation immunity. NMOS STI radiation immunity after gamma irradiation improved to the level of other test structures when boron body implant was skipped. Similar to [9], we interpret it by the deactivation of boron implant in Si substrate by hydrogen released from Si/SiO_2 interface by the ionizing radiation.

Skipping of the boron implant in the bulk of STI devices, identical to the sensing region of the C-Sensor, resulted in the increased radiation immunity.

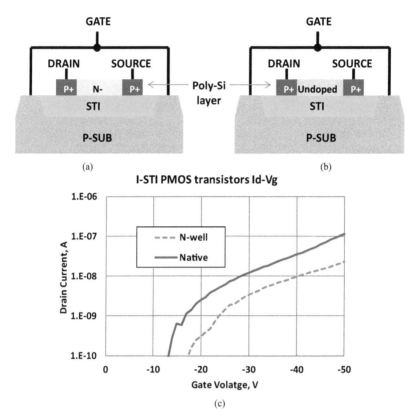

Figure 6.16 "Inversed" PMOS transistors with STI gate dielectric: (a) with N-doped "substrate"; (b) with undoped "substrate"; (c) typical Id–Vg curves. The lateral dimensions of I-STI PMOS devices and gate dielectric thickness are the same as those for C-Sensor control capacitor. Poly-Si doping dose is $\sim$0.5E 14 cm^{-2}, N-type.

This significantly broadens the range of measured radiation doses because the floating gate can be programmed and discharged by radiation several times without device degradation.

6.4.2 FG Devices for Monitoring the TID

6.4.2.1 Introduction

In direct semiconductor detectors, the absorbed radiation is converted into electrical signal due to direct interaction with the semiconductor material, without a need for the external converting techniques. Three well-known

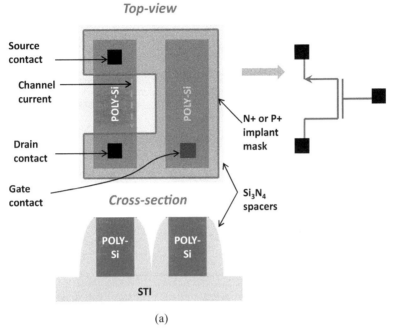

(a)

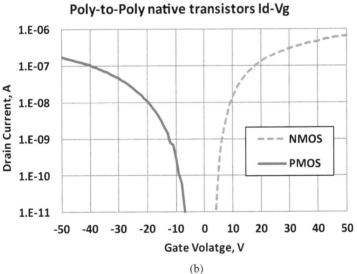

(b)

Figure 6.17 Poly-to-Poly "native" transistors: (a) structure diagram; (b) typical Id–Vg curves. Poly thickness (acting as the width of the transistor) is 0.25 μm, poly line spacing (the filling dielectric acting as transistor gate dielectric) is 0.3 μm. The length of undoped Poly region (transistor channel length) is 2 μm.

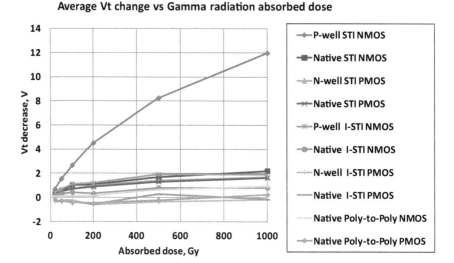

Figure 6.18 The average Vt change vs. Gamma radiation absorbed dose for all types of tested structures. All data are normalized, so that positive Vt changes correspond to positive trapped charge in the dielectrics.

types of direct semiconductor radiation sensors are based on (i) diode leakage [22–24], (ii) capacitor discharge [25] and (iii) MOS transistor degradation [26–30].

In the first case, the PIN diode is held at high reverse bias, while the leakage current is monitored (Figure 6.19). The ionizing agent (charged particle or photon) creates charge in the depletion region and contributes to the leakage current. The increase in leakage current is detected by the external circuitry, analyzed and then a conclusion on the ionizing agent parameters (energy, dose, etc.) can be made. The high volume of diode intrinsic region is required to increase the sensitivity of the diode-type sensor.

The operation of capacitor-type sensor (Figure 6.20) is also based on collection of charge produced by ionizing agent, but the method is different. The capacitor is charged before the exposure, and the charge, created by ionizing agent in capacitor dielectric, is separated by electric field and drifts to capacitor plates, leading to gradual discharge of capacitor. As a result, the capacitor voltage decreases and the information about ionizing agent can be obtained.

Another type of semiconductor radiation sensors utilizes the degradation of MOS transistor dielectric, caused by ionizing radiation. The radiation

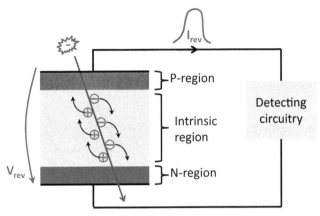

Figure 6.19 Direct semiconductor radiation sensor based on diode leakage.

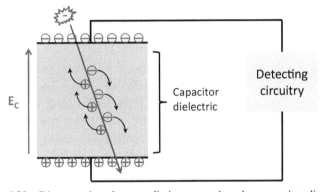

Figure 6.20 Direct semiconductor radiation sensor based on capacitor discharge.

creates traps (dangling bonds) in silicon oxide, which can be occupied by charges (Figure 6.21(a)). This, in turn, leads to changes in drain–source current of the transistor (Figure 6.21(b)), which can be detected by external circuitry. The drawback of this sensor type is that the trapped charges are not stable: they can de-trap and sensor readings will change unrelated to radiation.

In case of diode-based and capacitor-based sensors, the radiation degradation is also present and can introduce peculiarities in sensors readings. However, these devices can be engineered so that the degradation influence is minimized and precisely estimated, so that correction schemes can be introduced to significantly reduce the influence of degradation. On the contrary, the transistor-based sensor is based solely on degradation. Since 100% of its

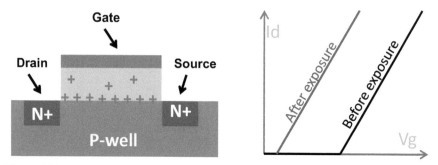

Figure 6.21   Direct semiconductor radiation sensor based on NMOS transistor degradation: (a) schematic cross section; (b) schematic demonstration of radiation influence on transistor Id–Vg curves.

signal comes from degradation, the performance of the sensor depends on its thermal history (annealing). Thus, the mistakes in dose estimation can be higher than in case of diode and capacitor, where only part of signal comes from degradation.

One of the ways to utilize capacitor principle for building radiation sensor is to use floating gate non-volatile memory device, which includes two or more coupled capacitors. The floating gate represents one of the capacitor plates. The FG is charged before the exposure and discharged by radiation. The radiation agent information can be extracted by measuring the current variation in the NVM device, as explained below.

6.4.2.2 C-Sensor operation principle

Floating gate devices are MOS structures with an additional conducting electrode electrically isolated from the gate and substrate. This electrode is called Floating Gate. The FG is surrounded by high-quality dielectric from all sides. There is no direct electric contact to it. The charge injected into the FG influences the Vt of the MOS transistor if the FG is a part of such a transistor. Two or more logic states can be represented by different Vt levels.

The FG devices found broad use as non-volatile memory cells in various electronic devices like memory cards, disc-on-keys, computer SSD drives and mobile phones.

Various types of FG NVM devices exist. In this section, we will consider FG NVM memory cell based on NMOS transistor (Figure 6.22(a)). Similar considerations can be used for PMOS transistor, as well.

A typical Id–Vg curve of NMOS transistors is shown in Figure 6.22(b). When FG is introduced into NMOS (Figure 6.22(c)), a similar Id–Vg curve can be measured (Figure 6.22(d)), while the gate is now called "Control Gate" (CG) to distinguish it from FG. Introducing FG allows to adjust Vt (i.e. to shift Id–Vg curve, Figure 6.22(d)) by injecting electrons or holes into the FG by utilizing Fowler–Nordheim tunnelling or one of the hot carrier injection mechanisms.

Following the above-described operation principle of capacitor-type radiation sensor, the FG NVM cell will be discharged by ionizing radiation, because it consists of two serially connected capacitors (FG-CG and FG-substrate). The detectors of ionizing radiation based on of-the-shelf NVM products were proposed [36–38]. They are pre-charged before the exposure and read failures during exposure are used for estimation of absorbed dose. The dose resolution is typically low, because the volume of sensing devices is small.

We utilized one of TowerJazz-embedded NVM solutions, called C-Flash [39–41], for sensing ionizing radiation. The design of the C-Flash device was modified to covert the NVM cell into a sensor. To enhance the sensitivity, the volume of control capacitor was increased (Figure 6.23). The control capacitor was formed using the STI dielectric. This dielectric has a thickness

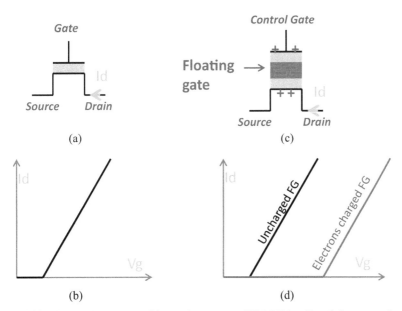

Figure 6.22 Converting (a) NMOS transistor to (c) FG NVM cell and demonstrating its operation: Id–Vg curves of (b) NMOS transistor and (d) FG NVM cell.

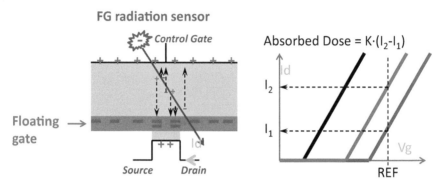

Figure 6.23   C-Sensor principle: (a) illustration of energetic particle detection and (b) the corresponding Id–Vg curves.

of 3500 Å, which is 30 times higher than gate oxide thickness (110 Å in 5 V CMOS technology).

We call the developed device "C-Sensor" [31–35] to distinguish it from its NVM prototype: C-Flash. The letter "C" stands for "complimentary", referring to CMOS inverter, which is used in the readout circuit of C-Flash and, at least, some of the C-Sensor flavours.

6.4.2.3 Implementation of C-Sensor principle in CMOS platform

As explained above, FG NVM cell can be utilized for sensing ionizing radiation. Nevertheless, to bring its sensitivity to a practical range, the sensing volume should be increased. The way to do it is shown in Figure 6.23(a). To implement such a structure in the standard CMOS flow, STI oxide was utilized to implement control capacitor, as schematically shown in Figures 6.24(a) and (b): the control capacitor is located adjacent to the sensing NMOS, sharing the floating gate made of polysilicon with the control capacitor.

The read-out principle of the sensor is the same as in the case of FG NVM cell. Charge injection into the floating gate by Fowler–Nordheim mechanism is utilized: high voltage is applied between the isolated P-well and the substrate (of NMOS).

The important characteristic of the radiation sensor is its coupling ratio (CR) – the relation between the control capacitor (STI) and read-out capacitor (GOX) values. A high coupling ratio (above 10) ensures high gm of the read-out NMOS (close to that of standard NMOS) and allows to minimize the injection voltages (the applied voltages falls on the GOX between the

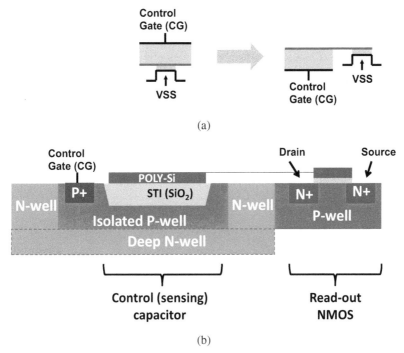

Figure 6.24 Implementation of C-sensor principle in the single-poly CMOS flow: (a) control capacitor arrangement; (b) schematic cross section.

isolated P-well and the substrate). The layout of sensor with minimum Design Rule (DR) NMOS and CR = 10 is shown in Figure 6.25. Read-out Id–Vg characteristics for different amount of charge in FG appear in Figure 6.26(a). The charging/discharging curves of the sensor are shown in Figure 6.26(b).

6.4.2.4 Detecting different types of ionizing radiation using C-Sensor

Typical discharge kinetics of C-Sensor (CLINAC gamma rays in this example) is presented in Figure 6.27. The C-Sensor is pre-charged to Vt = 4V and exposed to the radiation beam from a medical linear accelerator CLINAC 700C of Varian. The minimal and peak energies of photons are 2 MeV and 5 MeV, respectively. The Vt of sensor is measured at different absorbed doses (5 Gy, 10 Gy, 20 Gy, 50 Gy, 100 Gy, 150 Gy). The dose rate was of the order of 1 Gy/min.

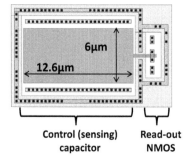

Figure 6.25 Layout of the radiation sensor. The layers "poly", "active", "contact" and "n-well" are shown.

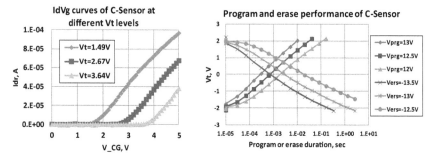

Figure 6.26 Electrical CZ of radiation sensor: (a) Id–Vcg curves for different amount of charge in FG; (b) charging/discharging curves (Vt vs pulse duration).

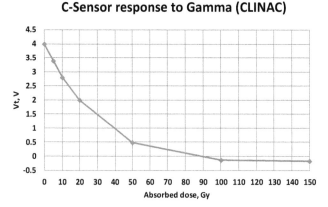

Figure 6.27 Typical kinetics of C-Sensor discharge.

As seen from Figure 6.27, at the beginning of discharge, the Vt loss is nearly linear (e.g. 99% linearity at the first 2 V of Vt loss). In this mode, the sensor is supposed to operate in the products. After continuous exposure, the Vt of the senor does not change any more with dose. In this research, we investigated the behaviour of different sensor flavours (different geometries and doping) in "linear" and "saturation" modes (the parentheses are used to avoid confusion with linear and saturation modes of MOS transistor operation).

To perform the statistical study of C-Sensor response to ionizing radiation, a matrix of 4096 and 2048 C-Sensors was developed. The irradiation experiments with arrays were performed in a manner similar to that of the single cells: the cells in arrays were pre-charged, irradiated by different doses and Vt distributions before and after exposure were recorded (Figure 6.28).

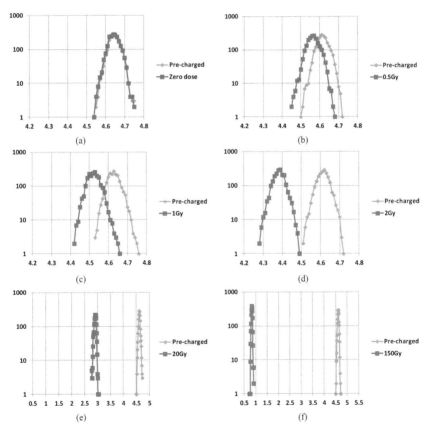

Figure 6.28 (b)–(f) Vm distributions of 2k arrays of sensors exposed to different doses: (a) the reference array (not exposed to radiation).

As in the case of single cells, we observe a nearly linear response to low doses (Figure 6.28(b)–(d)) – nearly 100 mV loss for 1 Gy dose. In the case of 20 Gy, the sensitivity decreases to ∼85 mV for 1 Gy, and the sensor is fully discharged by 150 Gy.

Another important observation from the array irradiation experiments is excellent sensor-to-sensor repeatability. This is a critical result for practical applications: it means that much smaller arrays can be used to build reliable dosimeter systems, while expected sensitivity could be at the level of 0.1 V/Gy.

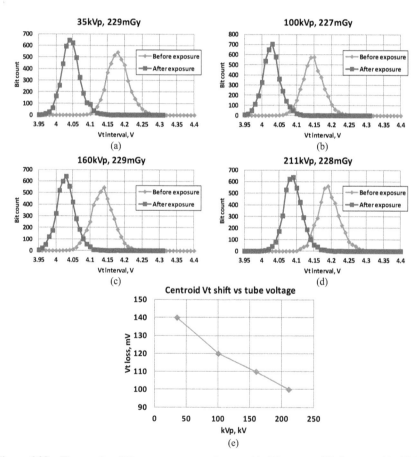

Figure 6.29   The results of X-ray energy experiment with 4K arrays of C-Sensors: (a)–(d) Vt distributions before and after exposures to X-rays of different energies; (e) dependence of Vt distribution peak shift on tube voltage.

As mentioned above, the demonstrated high repeatability of the array sensors allows to perform sensitivity enhancement by electronic means. Moreover, this high stability allowed proposing additional operation modes and new applications, like gamma imaging and spectroscopy and detection of energetic ions.

The dependence of C-Sensor response on X-ray photon energy was investigated using 4K matrices. All sensors in arrays were charged uniformly to Vt about 4V and exposed to 228±1 mGy dose of X-rays of different energies. After the exposure, the Vt array distributions were re-measured and the Vt distribution centroid shift was calculated for each energy (Figure 6.29).

A pronounced and strongly repeatable dependence of the sensor response on X-ray photon energy is observed.

The response of C-Sensor to alpha particles from Po-220 source (5MeV) was investigated. As in the above-mentioned irradiation experiments, the 4K matrices of C-sensors were charged to Vt = ~4 V and exposed to alpha source. From the experimental results, it appears that interaction of C-sensor with single alpha particle results in Vt decrease of about 1 mV (Figure 6.30). Thus, by using proper amplifier circuitry, single alpha particles can be detected.

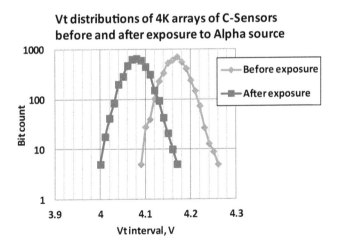

Figure 6.30   Response of C-Sensor to alpha particle.

6.5 Conclusion

In this chapter, challenges of CMOS technology radiation hardening from the process point of view have been presented with the focus on both core CMOS and embedded polysilicon floating-gate and charge-trap non-volatile memories. Physical phenomena responsible for radiation immunity were analysed and the corresponding process challenges and possible solutions were presented. Special test structures for the control of charge accumulation and interface degradation in employed dielectric films were designed that allowed to analyse the effects in CMOS front end under different types of radiation and make reliability predictions. Low power consuming FG sensors were introduced into rad hard process flows. Corresponding devices do not need additional masks and allow to combine the rad hard process technology and on the chip radiation dose measuring system. Since the FG radiation sensors are small and ultra-low power consuming, they are suitable for a variety of challenging applications, in particular when developing chips for sterilization systems, radiation safety, security devices, in-vivo dosimeters for precise dose delivery control in radiation therapy and other medical appliances. Availability of rad hard CMOS processes with the mentioned above features makes RHBD more efficient and allows to reach outstanding performance of both digital and analog products.

References

[1] A. Strum, T. Mahlen and Y. Roizin, "Non-volatile memories in the foundry business," in Memory Workshop (IMW), Seoul, South Korea, 16–19 May 2010.

[2] TSMC 2009 Brochure: Embedded non-volatile memory.

[3] W. Brown and J. Brewer, Nonvolatile semiconductor memory technology: a comprehensive guide to understanding and to using NVSM devices., IEEE Press, 1998.

[4] Y. Nishi, Advances in Non-volatile Memory and Storage Technology, Elsevier, 2014.

[5] C. Calligaro, U. Gatti, E. Pikhay, Y. Roizin, "Radiation-Hardening Methodologies for Flash ADC", Analog Integrated Circuits and Signal Processing Journal (Springer), May 2016, Vol. 87, page. 141–154.

[6] US Patent 9,431,455.

[7] R. Fuller, "Radiation hardened dielectric for EEPROM". US Patent 6,130,172.

[8] N. V. Shlopak, Y. A. Bumai and A. G. Ulyashin, "Hydrogen passivation of γ-induced radiation defects in n-type Si epilayers," Phys. Status Solidi A, vol. 137, no. 1, pp. 165–171, 1993.

[9] Y. Keqin, Y. Suge and L. Shijin, "A Study on High Density Gate-Oxide Anti-Fuse PROM Memory Cell," in ICMIE, 2016.

[10] J. Wang, "Radiation Effects in FPGAs," in Workshop on the Electronics for LHC Experiments, 2003.

[11] H. Jae, L. Hyunjin, K. Wansoo, Y. Gyuhan and Y. C. Woo, "On-State Resistance Instability of Antifuses during Read Operation," in The 21st Korean Conference on Semiconductors.

[12] US Patent 9,224,496.

[13] Kaczer, "Impact of MOSFET gate oxide breakdown on digital circuit operation and reliability", IEEE Trans. on Elec. Dev, vol. 49, no. 3, pp. 500–507, 2002.

[14] "RTG4 Radiation Update from MicroSemi using UMC platform," 2017.

[15] "Radioisotope Power Source Embedded in Electronic Devices". US Patent 20160379729.

[16] K. N, "Tritium-a MicroPower Source for the on-Chip Applications".

[17] Y. Roizin and E. Pikhay, "Photovoltaic device with lateral P-I-N light-sensitive diodes". US Patent 8,344,468, 1 January 2013.

[18] Y. Roizin, E. Pikhay, I. Chen-Zamero, O. Eli, M. Asscher and A. Saar, "Photovoltaic device formed on porous silicon isolation". US Patent 8,829,332, 9 September 2014.

[19] S. Gerardin and A. Paccagnella, "Present and Future Non-Volatile Memories for Space," IEEE Transactions on Nuclear Science, vol. 57, no. 6, 2010.

[20] P. McWhorter, S. Miller and T. Dellin, "Radiation response of SNOS nonvolatile transistors," IEEE Trans. Nucl. Sci., Vols. NS-33, no. 6, pp. 1414–1419, 1986.

[21] D. Adams, D. Mavisz, J. Murray and M. White, "SONOS nonvolatile semiconductor memories for space and military application," in IEEE Aerospace Conference Proceedings, 2001.

[22] N. Auricchioa, E. Carolia, A. Donatia, W. Dusib, P. Fougeresc and G. Landinia, "Thin CdTe detectors mounted in back-to-back confguration: spectroscopic performance for low-energy X- and gamma-rays," Radiation Measurements, no. 33, p. 867–872, 2001.

[23] "Silicon PIN Diode Radiation Detectors,," http://www.carroll-ramsey.com/detect.htm, Carroll-Ramsey Associates, Berkeley, CA, 1999.

[24] F. Oliveira, L.L.Amaral, A.M.Costa and T.G.Netto, "In vivo dosimetry with silicon diodesin total body irradiation," Radiation Physics and Chemistry, vol. 95, pp. 230–232, 2014.

[25] T. Jackson and L. Fine, "Dosimeter charging and/or reading apparatus". US Patent 4,224,522, 23 September 1980.

[26] A. B. Rosenfeld, S. Tavernier and et al., "Semiconductor Detectors in Radiation Medicine: Radiotherapy and Related Applications," in Radiation Detectors for Medical Applications, Springer, 2006, pp. 111–147.

[27] A. Rosenfeld, A. Holmes-Siedle and I. M. Cornelius, "Edge-on Face-to-Face MOSFET for Synchrotron Microbeam Dosimetry: MC modeling," IEEE Transactions on Nuclear Science, vol. 22, no. 6, pp. 2562–2569, December 2005.

[28] A. Holmes-Siedle and L. Adams, "RADFET: a review of the use of metal-oxide-silicon devices as integrating dosimeters," Radiation Physics and Chemistry, vol. 28, no. 2, pp. 235–244, 1986.

[29] G. Sarrabayrouse and S.Siskos, "Low dose measurement with thick gate oxide MOSFETs," Radiation Physics and Chemistry, vol. 81, p. 339–344, 2012.

[30] W.-C. Hsieh, H.-T. D. Lee, F.-C. Jong and S.-C. Wu, "Performance Improvement of Total Ionization Dose Radiation Sensor Devices Using Fluorine-Treated MOHOS," Sensors, vol. 16, no. 450, pp. 1–6, 2016.

[31] E. Pikhay, Y. Nemirovsky, M. Gutman, G. Villani and Y. Roizin, "Radiation Sensor Based on C-Flash Floating Gate Device," in Israel Vacuum Society, 2011.

[32] E. Pikhay, Y. Nemirovsky, M. Gutman, G. Villani and Y. Roizin, "Radiation sensor based on C-Flash floating gate device," TowerJazz Technology Journal, vol. 3, pp. 13–16, June 2012.

[33] E. Pikhay, Y. Nemirovsky, Y. Roizin and E. Villani, "Radiation array sensor based on C-Flash floating gate device," in SENSO, Gardanne, Aix-En-Provence, France, 2014.

[34] E. Pikhay, Y. Nemirovsky, Y. Roizin, V. Dayan, K. Lavrenkov, Y. Leibovich and D. Epstein, "Radiation sensor based on a floating gate device," in IEEE 27-th Convention of Electrical and Electronics Engineers in Israel, Eilat, 2012.

[35] E. Pikhay, Y. Roizin and Y. Nemirovsky, "Characterization of Single Poly Radiation Sensors," Electron Device Letters, vol. 36, no. 6, pp. 618–620, 2015.

[36] L. Z. Scheik, P. J. McNulty and D. R. Roth, "Dosimetry based on the erasure of floating gates in the natural radiation environments in space," IEEE Transactions on Nuclear Science, vol. 45, no. 6, 1998.

[37] T. Z. Fullem, L. P. Lehman and E. J. Cotts, "Examination of the utility of commercial-off-the-shelf memory devices as X-ray detectors," in 2007 IEEE Nuclear Science Symposium Conference Record.

[38] L. Z. Scheick, P. J. McNulty, D. R. Roth, M. Davis and B. E. Mason, "Measurements of dose with individual FAMOS transistors," IEEE Transactions on Nuclear Science, vol. 46, no. 6, pp. 1751–1756, 1999.

[39] Y. Roizin, E. Aloni, A. Birman, V. Dayan, A. Fenigstein, D. Nahmad, E. Pikhay and D. Zfira, "C-Flash: An Ultra-Low Power Single Poly Logic NVM," 2008 Joint Non-Volatile Semiconductor Memory Workshop and International Conference on Memory Technology and Design, pp. 90–92, 18–22 May 2008.

[40] H. Dagan, A. Shapira, A. Teman, A. Mordakhay, S. Jameson, E. Pikhay, V. Dayan, Y. Roizin, E. Socher and A. Fish, "A Low-Power Low-Cost 24 GHz RFID Tag With a C-Flash Based Embedded Memory," IEEE Journal of Solid-State Circuits, vol. 49, no. 9, pp. 1942–1957, 2014.

[41] H. Dagan, A. Teman, A. Fish, E. Pikhay, V. Dayan and Y. Roizin, "A Low-Cost Low-Power Non-Volatile Memory for RFID Applications," in IEEE ISCAS, 2012.

7

Rad-hard Resistive Memories

Cristiano Calligaro

RedCat Devices, Italy

"Survival is the ability to swim in strange water."

Frank Herbert, Dune

This chapter discusses some aspects related to RHBD of resistive memories (ReRAMs). The architectural and circuit solutions presented here need to fit the technological requirements related to the resistive oxide devices and the most effective hardening techniques.

First, some important architectural aspects for ReRAMs are examined and their main functional blocks are considered. For the most critical ones, such as column decoding and read-out block, suitable circuit solutions are provided. Then, a practical example of resistive memory in HfO_2 (Hafnium dioxide) is discussed.

7.1 ReRAM Cell

Resistive memories, generally called ReRAM (or OxRRAM, Oxide resistive RAM), belong to the category of Back-End-Of-Line (BEOL) memories. It means that the storage device is manufactured during the metallization steps, with a large reduction of masks and, consequently, of related costs. In some architectures, moreover, the achievable densities and power efficiency are high, so that resistive memories seem to be promising to become a concrete solution for non-volatile memories (NVMs) implementations.

However, in the scenario of commercial NVMs, it is well known that Flash technology still represents a winning solution for mass applications, thanks to its effectively low cost per bit, and a technological maturity that guarantees a high reliability. Nevertheless, resistive memories, even in this early stage of their technological life cycle, offer a good profitability in the space application scenario thanks to their intrinsic robustness to radiations due to the absence of any charge retention mechanism. Further market segments such as medical and automotive will pave the way as far as ReRAM becomes a more mature technology.

Indeed, the ReRAM principle is based on a non-volatile memory phenomenon realized by reversibly changing the resistance of a thin-film layer material sandwiched between two metal electrodes in a typical MIM (Metal–Insulator–Metal) structure. Generally, a broad range of inorganic and organic materials can be used for ReRAM, such as chalcogenides and metal oxides. Between the latter, there are the so-called n-type transition metal oxide compounds (i.e. the popular high-κ gate dielectric HfO_2, or TiO_2) and p-type transition metal oxide compounds (i.e. NiO). In this type of memory, the formation and transport of oxygen vacancies inside the oxide is the main physical mechanism, which leads to the resistance value change. The oxide itself shows an insulator behaviour (such as a capacitor), while defects introduced by oxygen vacancies create a conductive path across the oxide. Without applied voltage bias and at room temperature, the defect number is too small to create charge transport. However, when a sufficiently high electric field is applied, oxygen ions migrate toward the anode and create vacancy defects. To accommodate oxygen ions without preventing further vacancies movement, a Ti layer is added, achieving titanium oxidation. A cross section of the device is shown in Figure 7.1. As mentioned, the Ti layer between the HfO_2 and TiN is necessary to facilitate the redox reaction.

The creation of the first Conductive Filament (CF) is quite critical, since it depends on the voltage waveform applied in the forming phase [1]. Once the first CF is established, the successive evolution and buildup occur faster. In a bipolar resistive switching (BRS), when a reverse voltage is applied to the MIM stack, the CF can break. The oxygen ions oxidize back the Hf atoms and interrupt the conductive path [2], so that initial insulating behaviour is recovered.

In the design of a complete memory device to exploit the free "benefit" (intrinsic radiation resiliency) of resistive memories, the digital and analog blocks of the periphery need to be properly designed accordingly to the memristive device electrical requirements and following Radiation-Hardening-By-Design (RHBD) techniques.

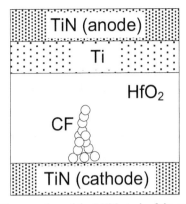

Figure 7.1   Cross section of the MIM stack of the resistive device.

In this chapter, some radiation-hardened architectures, based on ReRAM bit-cells, with the relevant functional blocks are presented. The other RHBD techniques already described in Chapters 2 and 3 (ELTs, EGRs, etc.) will not be repeated here.

7.2 ReRAM Array

Resistive memory arrays are generally designed according to basic architectural templates: crossbar and 1T1R. The crossbar approach, shown in Figure 7.2, is used when high densities are needed and has the great advantage to use the memory cell only without any selector [3]. Considering that the whole array is built between metals in the BEOL, the possibility to obtain very large arrays is concrete, but without a selector, word-lines and bit-lines, during the bit selection, may disturb neighbour cells. Many efforts are still ongoing in the crossbar direction but maturity both from the memory cell itself and from the crossbar approach is quite far and not easily implementable in real devices in short times [5,6]. Crossbar is still in the domain of research and development with relevant efforts from major players.

The second approach is the so-called 1T1R (see Figure 7.3), and it is very similar to other emerging memories and old fashion EEPROM. In this case, the selector is represented by a standard n-channel transistor connected to the MIM stack of the resistive cell. If the crossbar approach cannot isolate the single memory cell from other cells not involved in any reading or programming operation, in the case of 1T1R, the selector can prevent any attempt of false programming or read disturbance.

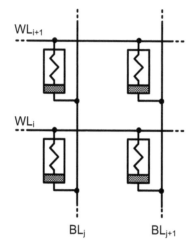

Figure 7.2   Crossbar resistive array.

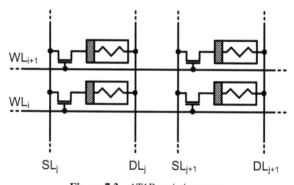

Figure 7.3   1T1R resistive array.

The penalty of 1T1R is evident and even if there is a clear benefit in making leverage on BEOL technology with the cell built between metal levels, nevertheless the presence of an n-channel transistor increases the overall area.

As mentioned above, the main benefit of an ReRAM cell is the intrinsic resistance against radiation coming from the absence in the MIM cell of silicon dioxide. In Horizon2020 R2RAM Project [4], a 1Mbit test vehicle has been irradiated with ^{60}Co up to 500 krad (Si) without any significant change in the current flowing in the memory cell. Of course, for the selector, RHBD must be used since, as a canonical n-channel transistor, it suffers from leakage currents along the borders of the channel (see Chapter 2).

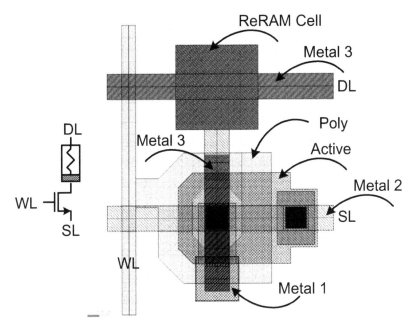

Figure 7.4   Physical view (layout) of the 1T1R bit-cell including the selector implemented with an ELT (250-nm BiCMOS technology).

To obtain a real rad-hard resistive cell, an ELT approach [7] is required in the selector as shown in Figure 7.4, where the drain of the selector is directly connected to the resistive cell with a direct connection to higher metal levels (BEOL).

The rad-hard resistive cell shown in Figure 7.4 (courtesy of IHP, Germany) is the real starting point to obtain a rad-hard complete memory [8]. In this cell (better discussed in Chapter 8), the HfO$_2$ layer is grown between metal 2 and metal 3, and the connection to the drain of the selector is guaranteed by a connection via Metal 1 in the opposite side of the cell itself to mitigate antenna effects during manufacturing. The three terminals of the cell are WL (word-line), SL (source-line) and DL (drain-line), which are available at the matrix level on poly, metal 2 and metal 3, respectively, leaving metal 1 on for internal connections inside the cell.

The overall result is a rad-hard cell having no issues from TID point of view thanks to ELT and not suffering from SELs thanks to the presence of n-channel transistors only (no SCRs in the array) and even if it is not optimized for the area it represents the best compromise between resiliency to radiations and area/power consumption.

Figure 7.5 shows a portion of a memory array built using the cell of Figure 7.4. Source lines and drain lines run with the same metal with which they are connected in the cell (Metal 2 and Metal 3) while word-line, originally in poly in the cell, is connected to a common Metal 1 line to reduce the distributed resistance that can introduce delays in the selection of the cell to be written or read.

Since the n-channel transistor acting as selector still has bulk to be connected, a biasing strategy must be adopted taking ground lines inside the array. Usually in standard memories, such lines can be connected every 16 or 32 cells, but considering the hostile environment represented by radiations even if SEL cannot be triggered (there are not p-channel transistors), a more frequent biasing connection is recommended. Every eight cells in a ground line in metal (Metal 1) is connected to the p-substrate of the array; in this way, any parasitic resistance coming from the substrate is avoided and the possibility to have a bulk connection of the selector to value higher than ground is null. Indeed, this is not an RHBD mitigation, but simply a precaution to guarantee a correct voltage distribution on cell terminals.

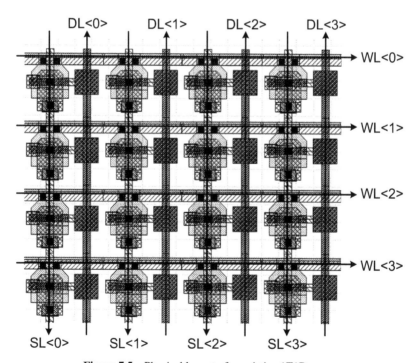

Figure 7.5 Physical layout of a resistive 1T1R array.

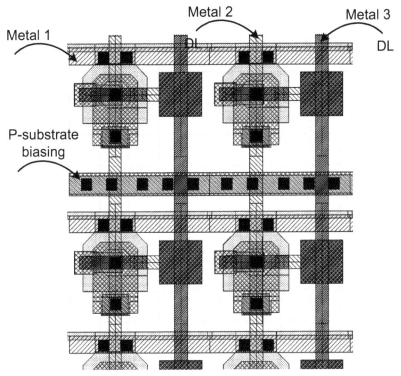

Figure 7.6 P-substrate biasing in the array.

As it will be better described in Chapter 8, programming and erasing a resistive memory cell is a quite complex operation requiring precise voltage and, when possible, applying increasing voltages starting from a specific value up to a maximum value like the Incremental Step Pulse with Verify Algorithm (ISPVA). To obtain voltage steps with precision in the order of tens of millivolts, it is crucial to have a stable bulk to avoid any noise coming from the substrate, thus affecting these algorithms (see Figure 7.6).

The resistive memory cell described in Figure 7.4, just like the major part of the cells belonging to this family, has different state conditions achievable by applying specific voltages, which, differently from traditional floating-gate memories, do not exceed power supply voltage.

When integrated for the first time, the cells are in a state called un-formed; this means that filaments have not been yet induced and the stack is a perfect MIM capacitor. Forming a resistive cell means creating for the first time a filament and it is important that it can be shaped in the best way because in the successive read and write operations, it will be a matter of dismantling and rebuilding filament on the original one.

Forming represents the first creation of the low resistive path and it shapes the behaviour of the cell in the successive operations. Since this operation takes the cell from a High Resistive State (HRS) to a Low Resistive State (LRS) implicitly, it is also a set operation but it is considered a wise approach using a specific (and possibly more accurate) algorithm when making a forming and use another one (possibly faster) when making a set of the cell. For these reasons, even if the target is the LRS, the physics behind is quite different and so requiring a different approach (see Chapter 8).

The operation of taking back the resistive cell from LRS to HRS is also called Reset; indeed, it does not recover the original un-formed condition since the original filament is only inhibited by interrupting the conductive path by removing vacancies. The result is that the HRS of an un-formed cell can in the order of MOhm while the HRS of a reset cell is in order of kOhm.

To summarize, we have the following working conditions:

- set/forming (creation/reconstruction of the low resistive path);
- reset (the bit-cell recovers its initial state) and
- read (reading phase)

From the data storage point of view, the cell states are:

- HRS (R_{OFF}) means logic state "1" and
- LRS (R_{ON}) means logic state "0"

The set/forming, reset and read operative bias conditions are summarized in Figure 7.7. The voltage V_{WL} controls the gate of the selector, while the driving voltage of drain (V_{DL}) and source (V_{SL}) can vary in the range of 1–3 V by applying the right algorithm.

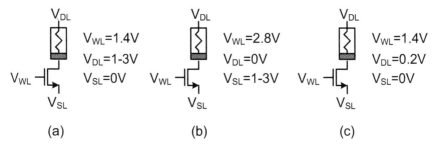

Figure 7.7   Biasing conditions for cell operations: (a) set/forming, (b) reset and (c) read.

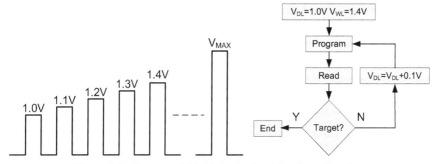

Figure 7.8 Program & Verify basic algorithm for cell setting.

Program & Verify (P&V) algorithms, as will be discussed in Chapter 8, optimize the low resistive state and high resistive state distributions. Moreover, as it happens for Flash memories, P&V algorithms are necessary to optimize the endurance of memory cells, as they minimize their electrical stress. P&V procedure consists in applying voltage pulses of increasing amplitude to a maximum value, and after each pulse, the resistive value of the cell is checked. When the target value is reached, the write (e.g. Set) operation is over and it is considered successful. According to the referred technology node, writing pulses amplitude starts at 1 V with an increase step of 100 mV, until a maximum voltage amplitude of 3 V is reached (see Figure 7.8).

The very basic algorithm shown in Figure 7.8 must be applied at bit level; this means that if a byte must be programmed (Set), then each bit must have its independent algorithm working since cells can reach the target in different times and with a different number of increasing pulses. Only when the last bit has reached the target, the programming operation can be considered over. When a resistive cell must be erased (Reset), the same algorithm will be applied keeping the control gate at 2.8 V and providing the SL with the same increasing train of pulses.

7.3 ReRAM Architecture

As extensively described in Chapter 2, a good floor-plan is the backbone of a rad-hard component, and the considerations described in Chapter 3 for SRAMs are valid also for non-volatile memory in general and resistive memories in particular. In the case of SRAMs, the building block was the 6T cell and the architectural solution adopted was substantially the independence of each bit among the byte. In this case, the same strategy can be used with an additional precaution already seen in Chapter 5: the differential cell.

In the differential cell approach, the single bit is stored in a pair of cells [11, 12]: in the first, an LRS is Set and in the second, an HRS is Reset. The "1" and the "0" will be discriminated according to the location of the HRS cell (left or right). The method is very simple and has the great advantage of not requiring a reference cell during reading operation.

Usually, non-volatile memories use as a reference an additional memory cell programmed in a very specific state. Historically, this comes from early EPROMs (Erasable Programmable Read Only Memories), where threshold voltage of memory cells was an issue in terms of variability at array level. Even by using very precise programming algorithm, the threshold of cell in a specific position could be very different from the threshold of another one because of the maturity of the process and even if all cells were conductive, currents were different, thus making discriminating the content difficult. To mitigate this effect, the solution adopted was the reference cell used as a comparison for the selected cell to be read. Indeed, in the EPROM era, the cell was not alone but shaped as a bit-line and this reference bit-line was repeated every 16 or 32 data bit-lines; in this way, the cell to be read was always close enough to the reference cell to avoid issues related to the variability of the process. Thus, the comparison was between the data cell having low threshold voltage and a reference cell never programmed or between a data cell having a high threshold voltage and the same reference cell never programmed. Sensing circuits were facilitated to discriminate data in an array suffering from process variability. With the progressive stability and maturity of CMOS process, reference bit-lines became single memory cells placed outside the array and duly programmed in fab when manufactured but still the approach of making a comparison between data cell and reference cell is a reality.

In rad-hard applications making use of reference cells, it is not advisable; the probability to hit the cornerstone of a sensing architecture is too risky and losing a reference cell means to lose the complete memory and not a single data bit that can be in some way corrected or substituted by more (or less) complex ECC engine.

The use of a differential cell in a non-volatile memory is an RHBD technique at architectural level and it impacts both the array and the periphery of the device. It is somehow radical but has the great advantage of completely avoiding the reference cell providing a read margin window as wide as the one in SRAMs without any issue related to bistable elements and critical charges. A differential non-volatile memory cell will never suffer from SEU simply because there is not any positive feedback loop triggerable by charged particles having a specific threshold LET.

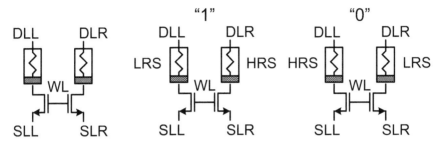

Figure 7.9 Differential resistive cell and definition of "1" and "0".

Figure 7.9 shows the differential resistive cell where it is possible to see how the pair shares the same word-line and two pairs of drain-line and source-line are discriminated to be Left (L) or Right (R). Conventionally, the bit is "1" if the LRS is in the left cell and "0" if in the right cell. Of course, HRS is in the complementary cell.

But from the RHBD point of view, there is still an open issue: if the cells pair, as depicted in Figure 7.9, is hit by a charged particle, the same sensitive volume of both cells can be involved, thus leading to a transient able to provide a wrong sensing, but if the cell are far enough from each other, then the same impinging charged particle cannot involve both cells. To do this, the safer is approach is to completely split the array into a Left Array containing all left cells and a Right Array containing all the right cells. This matrix split can be easily driven by decoding schemes able to reconstruct the virtual pair in a single cell having half bit on one side and the other half far enough to guarantee immunity to transients.

Figure 7.10 shows the architecture of a single matrix array making use of a resistive differential cell. The matrix is split into left and right using a mirroring approach, thus meaning that the position of each resistive cell of the pair is at the same distance from the terminal driver of the world-line so minimizing the mismatch coming from two elements manufactured in two different parts of the array. Row drivers serving left and right arrays are placed in the middle to make the routing from Row DEMUX easier, which is obviously in common and column drivers are specialized for drain-lines and source-lines (for programming and erasing purposes). Column DEMUX can be the same as the Row one. Figure 7.10 does not show the engine devoted to implement P&V algorithm requiring, as a matter of fact, an FSM to be handled according to RHBD techniques.

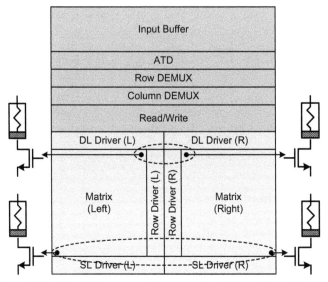

Figure 7.10   Single Bit array architecture using differential resistive cell.

The approach of matrix splitting can be extended to a complete device, as shown in Figure 7.11, where a similar approach to the one described in Chapter 3 for SRAMs is adopted to reduce MBUs to SEUs only [10–13]. In this case, eight independent blocks are replicated, each having its own decoding systems and leaving in common only input buffers and ATD engine including P&V algorithm that, if carefully designed at circuit level, can be rad-hard enough to avoid MBUs during programming, erasing or reading.

The penalty to be paid in such architecture is mainly related to area; doubling the matrix array and replicating eight times the decoding schemes can lead to very large chips.

Just like SRAMs, it is a matter to understand where the device is expected to be used (which applications) and if it is a stand-alone or an embedded device. For stand-alone devices, the architecture of Figure 7.11 can be an issue so at least a reduction of DEMUXs common blocks serving the whole memory is recommended as shown in Figure 7.12. In this way, a common peripheral engine implementing decoders, input buffers, ATD (if the memory is asynchronous) and P&V engine is used for independent split matrix arrays.

Considerations done for SRAMs are also valid here: if a charged particle hits a row/column driver, it affects only half bit, and in case of wrong sensing, there is only an SEU. On the other hand, if the charged particle hits a common

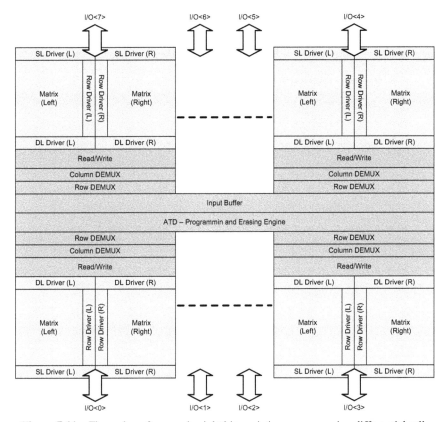

Figure 7.11 Floor-plan of a generic eight bits resistive memory using differential cell.

peripheral DEMUX, a complete word or column decoding can occur, thus causing an MBU.

For embedded resistive memories, things can be different and hybrid solutions between Figures 7.11 and 7.12 can be considered according to the requirements coming from the final applications and radiation environment (LEO, GEO, HEO).

7.4 ReRAM Periphery

Figure 7.13 shows the schematic of a row terminal driver of complete 1 Mbit (2 Mcell) resistive memory integrated in a standard 250 nm CMOS Process (by IHP). Differently from the terminal driver shown in Chapter 3 for SRAMs, in this case, the word-line has 256 cells (selectors) in common,

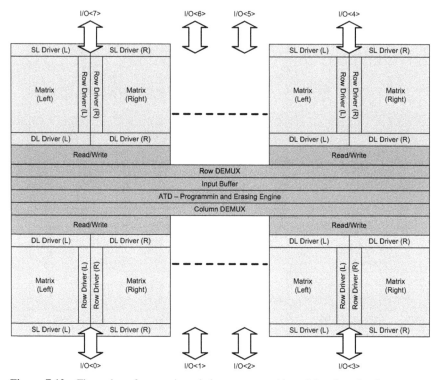

Figure 7.12   Floor-plan of a generic resistive memory with peripheral engine in common.

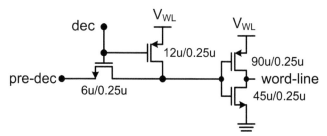

Figure 7.13   Row driver in a stand-alone resistive memory (H2020 R2RAM Project).

so the distributed capacitance (also considering the technological node) is relevant and the driver must have enough strength to access the farthest cell in the array with reasonable times.

The final inverter, driven by a p-channel boosting the input to V_{WL} and an n-channel acting as switch, has a wide width to enable high fan-out. V_{WL}, as mentioned above, can be 1.4 V in read mode and during programming, while it must be 2.8 V during erasing.

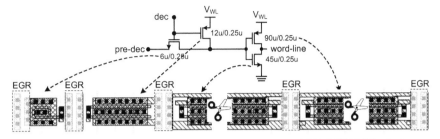

Figure 7.14 Row driver layout in a resistive memory (H2020 R2RAM Project).

Figure 7.14 shows the physical layout where p-channel and n-channel of the final inverter are shaped with long ELTs to keep the pitch of the resistive array. It is worth underlining the extensive use of EGRs to avoid SELs in regions where n-channel transistors and p-channel transistors are facing each other.

Concerning column terminal drivers, a possible strategy is to adopt a two-level scheme, where the first level (Y_M) is the one getting in connection the bit-line directly to sense amplifier for reading or programming/erasing, while the second gets the direct connection to the drain line (DL) or the source line (SL) according to the operation to be performed. Figure 7.15 shows the basic scheme for the first level ($Y_M<0:j>$), which is an n-channel switch connected to an array of switches ($Y_{ND}<0:k>$) gaining access to the drain-line and another array of switches ($Y_{NS}<0:k>$) gaining access to the source. DEMUXs must of course provide not only the selection of the cell to be accessed but also the direction (from drain to source or vice versa), and once the direction is selected, the counter-part (drain-line or source-line) must be grounded according to the operation conditions described in Figure 7.7.

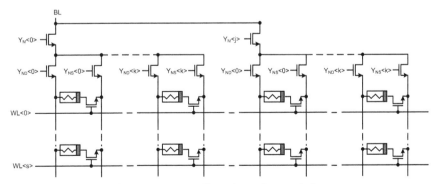

Figure 7.15 Column decoding scheme for a resistive memory.

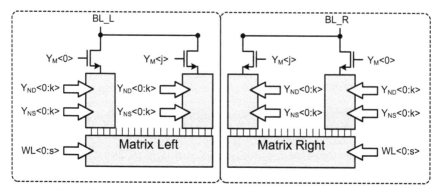

Figure 7.16   Column decoding scheme for left and right matrix arrays.

Since the matrix is mirrored, the decoding scheme must be replicated as shown in Figure 7.16; in this way, bit-line will become bit-line-left and bit-line-right, thus taking to an architecture very similar to the one used in SRAMs, where bit-lines belong to the bistable element represented by the inverter pair connected to the output of the first to the input of the second and vice versa.

Figure 7.17 shows the physical layout of the decoding scheme described above, where one Y_M terminal driver is connected to 16 Y_N drivers for both drain-lines and source-lines. All drivers (n-channel transistors) are ELT shaped to avoid channel degradation; this precaution is mandatory in this case, since in 250 nm CMOS process, silicon dioxides are thicker in more scaled technology (e.g. 180 nm) and degradation of channel borders is an issue.

Since no p-channel transistors are involved, the risk of SELs is null; nevertheless, a p-substrate biasing ring is placed all around the terminal stage block to isolate from external peripheral parts (DEMUXs).

As mentioned above, word-lines, connected to the gate of cell selector, must have two working voltages: 1.4 V during reading and programming and 2.8 V during erasing. This means that row terminal driver (see Figure 7.13) must be driven by a DEMUX, not only making the selection of the terminal stage, but also providing the right current, and by considering that the overall logic can be at a different power supply voltage (e.g. 2.5 V), it is evident that a level shifter is required making leverage on a voltage regulator (if the word-line must be connected to a voltage lower than power supply) or a charge-pump (if the voltage to be applied is higher than power supply). Of course, if a high voltage external pin is available (e.g. V_{PP}) in this case, it is only a matter of making the level shifter work between the required voltages.

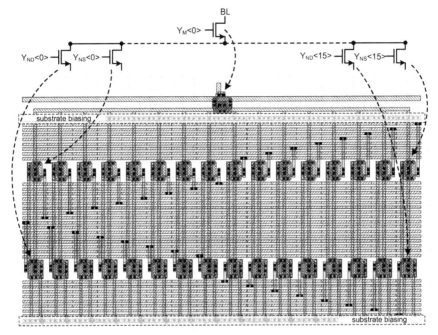

Figure 7.17 Physical layout of a column decoding sub-block 1–16 (Horizon2020 R2RAM).

Since row decoding, according to the terminal driver of Figure 7.13, is done using a pre-decoding (also known as 1st level decoding) and a fine-decoder (2nd level decoding), both signals must be boosted to high or low voltage to drive properly the terminal stage. Figure 7.18 shows the basic scheme where two level shifters boost the output: if the input signal is "0", it remains as is, but if it is at "1" (and the "1" is V_{DD}), it is taken to V_{WL}. In cascade to level shifters, logic elements (buffers or inverters) working at the same V_{WL} may square the signal before driving the terminal driver. The result is a driver having at the inputs logic signal already working at the right voltage and a buffer powered at V_{WL} giving strength to the word-line.

Radiation hardening in level shifters and all circuits handling different voltages (below or above power supply) is particularly insidious because of SELs.

Figure 7.19 shows an excerpt of the level shifter with a buffer in cascade coming from the scheme described in Figure 7.18; the layout in this case must carefully consider all possible SCRs, not only because there are p-channel transistors and n-channel transistors faced each to the others but also because

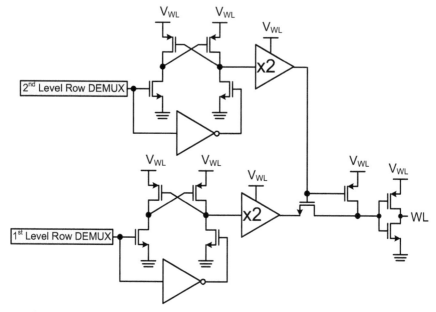

Figure 7.18   High- and low-voltage distributions to word-line using level shifters.

power supply is different from the one used in the rest of the chip. The level shifter is closed together with the buffer in a pair of guard rings: the first, connected to ground, enclosing all n-channel transistors and the second, connected to local power supply, enclosing p-channel. Where the two guard rings are faced, there is the EGR guaranteeing the cut of the positive feedback loop of the parasitic SCR.

When this block is placed in the layout close to another one having a different power supply (e.g. V_{DD}), some precautions must be considered to avoid again SELs. Figure 7.20 shows the same block of Figure 7.19 placed in proximity of the logic DEMUX working at a different power supply. Also in the case of the DEMUX n-channel and p-channel, transistors are enclosed in their own guard rings and the two blocks (Level Shifter and DEMUX) are placed with power supplies in opposite directions leaving, as shown in Figure 7.20, the common ground in the middle. This simple precaution avoids any SEL, since from any point of view, parasitic SCRs cannot be triggered, not even between two different power supplies.

As mentioned above (and better described in Chapter 8), resistive cells, during read mode, must be kept under a specific condition to avoid stress on the HfO_2 layer and to guarantee the right current flowing in the filaments. Of

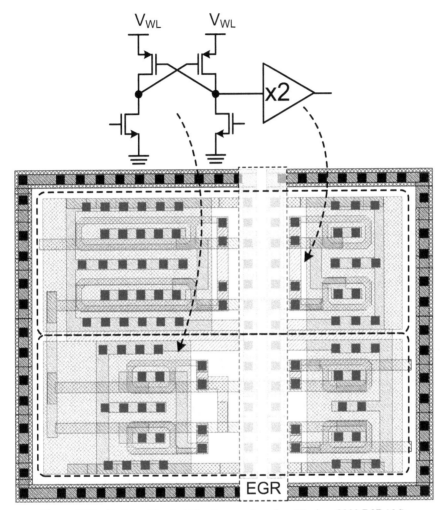

Figure 7.19 Rad-hard level shifter for row decoding (Horizon2020 R2RAM).

course, this current must be high enough to guarantee an efficient sensing, thus taking into account the classical trade-off between stress and sense: the lower the former, the more complex the latter and vice versa.

According to the material used in the MIM layer, drain-line voltage may vary. In Horizon2020 R2RAM Project (in IHP 250 nm process), this value has been below 0.5 V and the solution adopted to keep bit-lines at this value is shown in Figure 7.21, where an operational amplifier driving an n-channel transistor keeps, thanks to the virtual ground, the bit-line around the value of V_{BIAS}. The same approach is used both for left and right arrays.

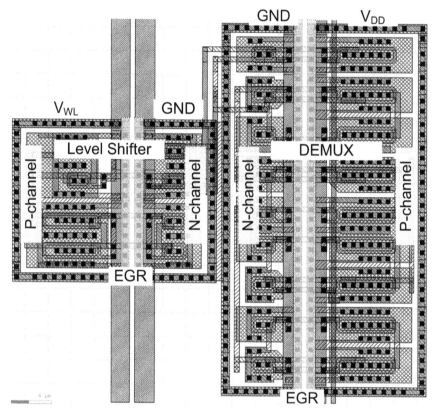

Figure 7.20   Placement of blocks working at different voltages (Horizon2020 R2RAM).

Figure 7.21 also shows the boundary conditions required to make a reading operation: Y_N decoding (2nd level decoding) enable the passage of the cell current through the drain-line of the cell while source-lines are moved to ground simply connecting the whole line passing through an enable (EN_{READ}) switch. Keeping Y_{NS} to ground (for both left and right) prevents any current passage through source and, meanwhile, keeping EN_{WRITE} to ground forces the cell current to pass only through the selected cell.

Once the resistive cells are selected, according to the differential approach, one will be HRS and the other will be LRS; this means that I_{CELL} will be very low in the branch where the resistance is high and vice versa in the other branch, where the current will be at the maximum value offered by resistive technology. Making a discrimination between a very low current and a higher one can be easily done with the circuit shown in Figure 7.22.

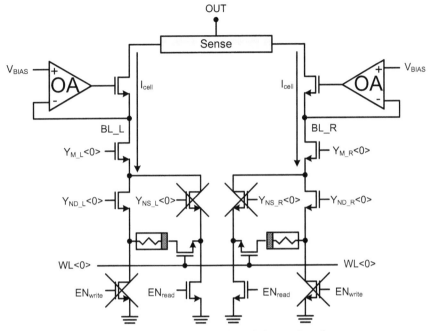

Figure 7.21 Cell biasing in resistive read mode.

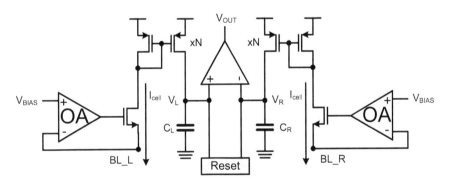

Figure 7.22 Sensing scheme using current cell mirroring.

The cell currents are mirrored via diode-connected p-channel transistors with a gain factor (N) to make the charge of the capacitors connected to the mirrors work more efficiently; in this way, along the branch having a higher I_{CELL}, the capacitor will be charged in a time shorter than the counterpart. A simple comparator connected to the capacitors can so discriminate the differential cell content by simply measuring the currents flowing. Once finished the sensing operation, a reset circuit can stop the reading operation and, meanwhile, by discharging the capacitor, it must take them back to the initial conditions for a new read.

Figure 7.23 shows the layout of part of the sensing scheme without comparator (which is very similar to those described in Chapter 3 for SRAMs). In this case, it is possible to see how C_L and C_R have been integrated by using

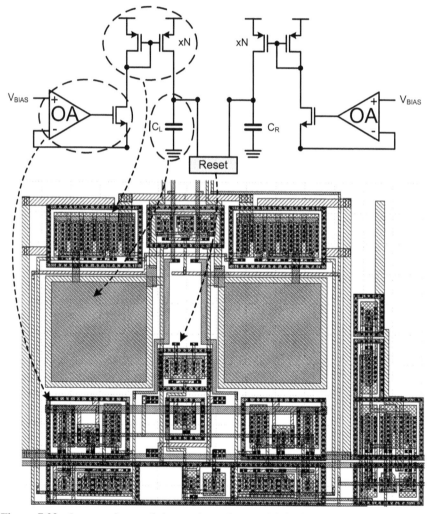

Figure 7.23   Layout of part of the sensing scheme with bit-line regulation and current mirrors.

MIM capacitors, which are particularly suitable for rad-hard components if considering that they are completely immune to radiations. If MIM capacitors are large enough to not suffer energy release modifying critical charges, then this solution can be considered radiation proof. The remaining structures (mirrors, voltage regulators and reset circuit) follow RHBD techniques extensively described in the previous chapters.

Concerning capacitors, even if charging times can be slightly different nevertheless at the end both, capacitors are charged and in that case, the comparator cannot discriminate any more the cell content. To avoid this problem, there are two options: the first is to use phases (generated by an ATD or coming from a clock if the resistive memory is synchronous) and the second is to build a Charge & Split system [14].

In the first case, a time window must be open to intercept where the charging curves are wider and when they start to be close to each other, stop the comparison and save the result of the comparator.

Figure 7.22 shows a possible solution using a Flip-Flop enabled to receive comparator output in a time window Δt.

From RHBD point of view, the solution of Figure 7.24 is safe enough if a good ATD or a synchronous system is adopted, making a filter on SETs; the only weak point, apart from the concept of charging capacitors C_L and C_R, which is intrinsic in the overall methodology, is represented by the Flip-Flop and the time window because if a charged particle hits during the time Δt, the comparator cannot recover in the time frame and what initially started as SET can become an SEU (and stored in the Flip-Flop). In any case, the probability to have an ion strike in a very short time frame (no more than 10 ns) is very low and if the comparator is duly designed using RHBD techniques, increasing critical charges in the most sensitive nodes, the overall scheme is safe enough.

To avoid the use of a Flip-Flop driven by a phase, a different scheme can be adopted as shown in Figure 7.25. This sensing scheme is also called Charge & Split and it can rely on a pair of n-channel transistors connecting the

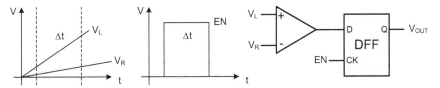

Figure 7.24 Sensing scheme based on latching in a time window.

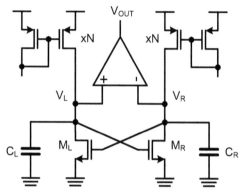

Figure 7.25   Sensing scheme based on Charge & Split methodology.

gate of the first with the drain of the second and vice versa. When capacitors C_L and C_R start to be charged by cell currents mirrored in the left and right branches, both gates of M_L and M_R are below the threshold voltage; since the two ramps have a different slope, one of the two gates will reach and overtake the threshold voltage before the other and by switching on the transistor, the effect is that it will start making flowing current further slow down the capacitance charge; the effect is that the faster ramp will reach its asymptotic voltage value as expected while the slower ramp will be inhibited.

Charge & Split sensing, originally conceived for multilevel Flash memories, is also suitable for rad-hard memories thanks to its intrinsic robustness and sensing precision (few μA can be discriminated). Since this structure does not need to a flip-flop, there is not the problem of a charged particle hitting inside the latching phase; once voltages V_L and V_R are split even with a transient, there is no possibility to flip the bit. Of course, this structure requires some RHBD precautions: 1) C_L and C_R must be large enough to not suffer energy release from charged particles and 2) transistors M_L and M_R must be ELT to completely avoid any degradation of the channel.

7.5 Forming, Setting and Resetting a Resistive Memory

In the previous paragraph, reading techniques have been described, but the most complex part of a non-volatile memory periphery is programming and erasing; in resistive memories, this feature has some additional issues to be considered.

As mentioned in Section 7.3, resistive cells must be formed when programmed for the first time, and considering the double array, a preliminary operation on all the cells is required. This can be done applying a P&V algorithm but if for the programming operation it is a matter of applying an increasing pulse series, on the other hand, verification cannot be done on a pair of cells both starting with a null current (HRS) and intended to be formed (LRS). In these conditions, the cells are identical, and the sensing schemes described in the previous paragraph are useless.

It is clear that the arrays must be considered independent and forming operation requires an additional reference to be used only for programming (and erasing purposes) and this reference cannot be another cell but another element emulating the behaviour of the resistive cell. This can be resistors selected to have a resistance equal to the one expected by the cell itself in LRS and HRS. Figure 7.26 shows the reference scheme making use of such strategy. R_R and R_S can be poly resistors, very stable under irradiation in almost all CMOS technologies and the reference chain must be the same as in the matrix so a first level of column decoding Y_{MR}, a second level of decoding Y_{NR} and a selector having the same shape of the resistive cell selector. Instead of the MIM, poly resistors R_R and R_S must be connected to their selector and they are enabled when verifying for Setting (or Forming) or Resetting. This scheme is intended to match the situation in the matrix array (at least from the dynamic point of view) and optionally, since many Y_N are

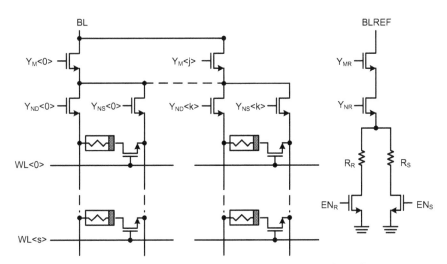

Figure 7.26   Reference scheme for forming, setting and resetting.

connected to Y_M in the matrix, a capacitance may be added to emulate the load coming from the matrix. The same approach is valid at bit-line level and in parallel to R_R and R_S, a capacitor emulating the bit-line load of the matrix counter-part may be added.

Under these premises, the column decoding scheme of Figure 7.16 can be changed as shown in Figure 7.27, where both the left and the right matrixes have their own reference cell to be used. Since the sensing scheme may be the same, two reading modules can be used: a) one for the standard read mode when differential cell is checked to see if HRS is on the left or on the right and b) two sensing schemes for left and right matrixes, each interfacing with their own poly reference resistors. We call Read Mode the first and Verify Mode the latter.

It is worth underlining that a complete resistive device can have eight (or more) bits so the scheme of Figure 7.27 must be replicated, thus leading to 24 sense amplifiers: 8 used in Read Mode and 16 in Verify Mode independently.

According to Figures 7.27 and 7.28, it is possible to define all working conditions and, in particular, define when the memory is in Read Mode (WE = 1 and OE = 0) and Verify Mode (WE = 0 and OE = 0) discriminating with SR if it is for Set/Forming or Reset. Since forming, setting and resetting operations involve arrays independently, and the signal AE tells if the operation is done on the left or on the right. CK is the clock in case an FSM is used to implement the P&V algorithm.

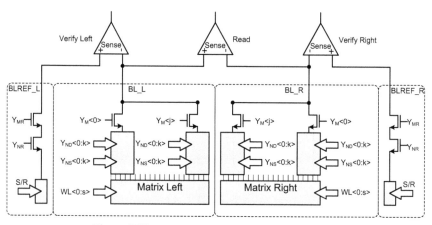

Figure 7.27   Read and verify in a resistive memory.

Operation	CE	WE	OE	SR	AE
Stand-by	1	X	X	X	X
Read Mode	0	1	0	X	X
Set/Forming (Array Left)	0	0	1	0	0
Reset (Array Left)	0	0	1	1	0
Verify Set (Array Left)	0	0	0	0	0
Verify Reset (Array Left)	0	0	0	1	0
Set/Forming (Array Right)	0	0	1	0	1
Reset (Array Right)	0	0	1	1	1
Verify Set (Array Right)	0	0	0	0	1
Verify Reset (Array Right)	0	0	0	1	1

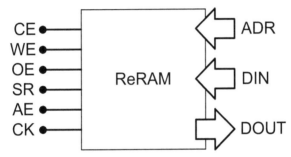

Figure 7.28   Truth table of a generic resistive memory.

Figure 7.29 shows the flowchart of a generic resistive memory related to forming the device. Once selected the address and the array (left or right) where to carry out the setting operation, the voltage to be applied to the drain-lines is set to the initial value (e.g. 1 V). This voltage may come from a charge pump or provided externally passing through a specific pin; in case of a charge pump, an FSM driven by a clock can make the engine self-consistent; otherwise, the overall set operation must be synchronized externally by a pattern generator or a programming memory machine.

The pulse is applied by moving to "0" WE, and after a delay time (10 µs is the typical half-period for this kind of memory), also OE can be moved to "0" to enter in the Verify Set Mode. Data out can be now evaluated and if all bits are formed, the memory can be taken in stand-by or reactivated for a new address; otherwise, if only part of the bits are formed, they are removed by the hits of the pulsing voltage. For bits not already formed, pulsing voltage is increased and a new Set/Verify Set cycle is performed until all bits have reached their target.

Operation	B0	B1	B2	B3	B4	B5	B6	B7
Current Data	0	0	0	0	0	0	0	0
Previous Data	X	X	X	X	X	X	X	X
Array Left	Set	Set	Set	Set	Set	Set	Set	Set
Array Right	Set	Set	Set	Set	Set	Set	Set	Set

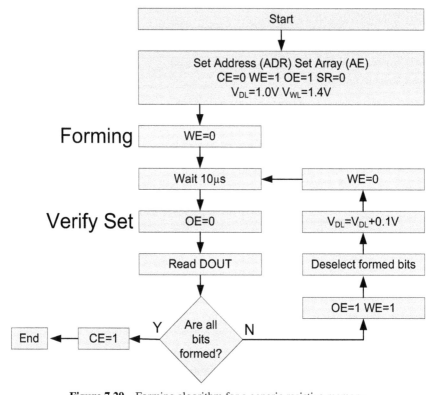

Figure 7.29   Forming algorithm for a generic resistive memory.

The algorithm of Figure 7.29 can be (and must be) fine-tuned for a resistive memory in the beginning if its life and all address must be scanned on both memory arrays. The final goal is to have all the memory cells in LRS and so ready for the first programming. It is worth underlining that a formed memory still has all bits in an undefined state since both left and right cells have the same state where a bit data is foreseen having one bit in HRS and one bit in LRS. To avoid this problem and make easier the algorithm to reprogram the memory, a second operation, after the forming (or as part of forming), is to write all "0" in the device or write a checkerboard. The goal is not to write something specific but to take all cell pairs in a defined state (one to HRS and one to LRS).

Figure 7.30 shows a possible algorithm for making the first write in a resistive memory just formed. In this case, all "0" must be written and considering that the left matrix is already in LRS, only the right array must be considered. Starting from the first address (the memory must be scan in all addresses), the Reset Mode is applied (SR = 1) and the Right Array is considered (AE = 1).

Operation	B0	B1	B2	B3	B4	B5	B6	B7
Current Data	0	0	0	0	0	0	0	0
Previous Data	X	X	X	X	X	X	X	X
Array Left	Set	Set	Set	Set	Set	Set	Set	Set
Array Right	Reset	Reset	Reset	Reset	Reset	Reset	Reset	Reset

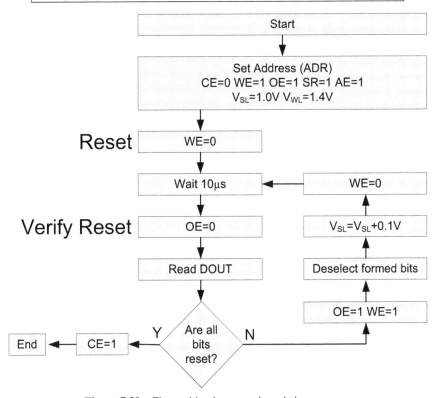

Figure 7.30   First writing in a generic resistive memory.

The rest of the algorithm is, at this point, the same of the one described in Figure 7.29 and a progressively increasing pulse series is applied to the source-lines of the cells achieving the passage from LRS to HRS.

The algorithms shown in Figures 7.29 and 7.30 are intended to be applied to the complete memory and, indeed, they can be part of the preliminary check of the chip at wafer level. In fact, some of the cells can be damaged or resistive layer growing can have problems in the forming operation. This is the typical check to be done by using probe-cards. If all cells are formed correctly and the first writing has been carried out correctly, the device can be considered functional and then moved to sawing, dicing and assembly in the right package.

If a resistive memory has been already formed and written for the first time, the algorithm to carry out a generic programming of a byte on a previously written one can be designed according to the generic flow of Figure 7.31. Regardless of the byte to be written, there will be always setting and resetting operations to be done on the left and the right arrays. And considering that passing from set to reset mode means accessing the cell from drain or from source, the best way is to set all cells needing to be led to LRS and vice versa in two different times. The flow of Figure 7.31 foresees the acquisition of the address and the byte to be written and a standard read of the byte content. Then, the algorithm checks the bits not needing to be changed, while for those needing a change, they are divided into two groups: the cell to be moved from LRS to HRS and vice versa. This operation involves both left and right arrays.

Since Verify Mode (both Set and Reset) uses data-out (DOUT) bus to evaluate if cells have reached their target, the operations described above must be done separately: 1) Set for Left Array, 2) Set for Right Array, 3) Reset for

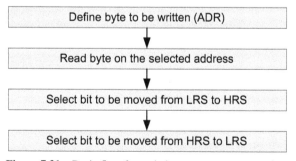

Figure 7.31   Basic flow for resistive memory programming.

Left Array and 4) Reset for Right Array. Considering that setting and resetting operations must be done not on all cells, some registers are required to mark, which are the cells to be set or reset in the specific array.

Figure 7.32 shows a flow for making such operations. From actual data and the previous data belonging to the selected address, the cells of the two arrays requiring to be moved to HRS or LRS are defined and a register is filled with "1" for cell to be set or reset and with "0" for cell that do not have to be moved from their state. In this way, the Verify mode (using DOUT) will simply have to make a comparison with the corresponding register and to apply the pulsing voltage only on those cells requiring to be programmed and discarding those who have already reached the target.

Figure 7.33 shows the flowchart of the algorithm used in the 1Mbit (2Mcells) RC27F1024R2RAM1 test vehicle developed in Horizon2020 R2RAM project. The algorithm (including increasing pulses on V_{DL} and V_{SL}) is applied externally by a testing board implementing a Xilinx Spartan3 FPGA on an independent mother board and a daughter board mounting the resistive memory only as Device Under Test (DUT). The two boards are connected via a flat cable to guarantee the possibility to conduct tests under irradiation and protecting the mother board (see Figure 7.34), which is not rad-hard.

Operation	B0	B1	B2	B3	B4	B5	B6	B7
Current Data	0	1	0	1	0	1	0	1
Previous Data	1	1	1	0	0	1	1	0
Array Left	Reset	X	Reset	Set	X	X	Reset	Set
Array Right	Set	X	Set	Reset	X	X	Set	Reset
Reset Byte AL	1	0	1	0	0	0	1	0
Reset Byte AR	0	0	0	1	0	0	0	1
Set Byte AL	0	0	0	1	0	0	0	1
Set Byte AR	1	0	1	0	0	0	1	0

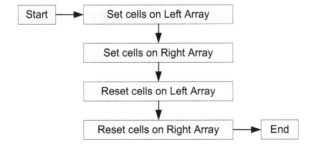

Figure 7.32 Re-programmability flow for a generic resistive memory.

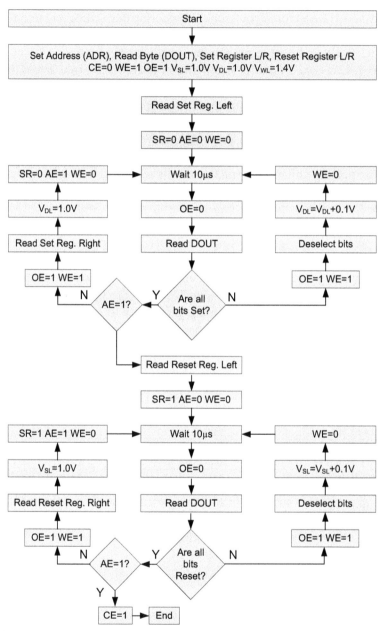

Figure 7.33   Programming flow for RC27F1024R2RAM1 (Horizon2020 R2RAM).

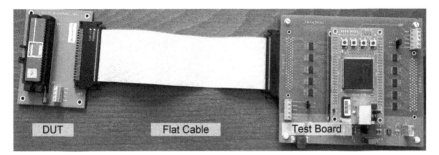

Figure 7.34 ReRAM Test Board using FPGA (Horizon2020 R2RAM).

Figure 7.35 shows the same board mounted in an experimental setup for a testing under heavy ions at RADEF in Jyvaskyla (Finland) and under ^{60}Co at University of Palermo (Italy). Both setup arrangements protect the mother board containing the FPGA, leaving exposed the daughter board, only mounting, on a ZIF socket, the resistive memory [15].

The approach for testing the resistive memory under irradiation considers the "in situ" methodology; in this case, the mother board driving the DUT is active and a personal computer (possibly protected in a different chamber)

Heavy Ions (RADEF, Jyvaskyla - Finland)   Gamma Rays (UniPA, Palermo - Italy)

Figure 7.35 ReRAM testing with Heavy Ions and Gamma Rays (Horizon2020 R2RAM).

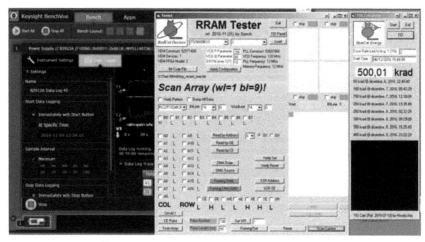

Figure 7.36   In situ ^{60}Co irradiation test on RC27F1024R2RAM1 (Horizon2020 R2RAM).

is connected via USB to the mother board. The PC itself is connected via the Ethernet cable and simply by using a standard SSH connection controlled remotely by an external (and more protected) chamber [15].

Figure 7.36 shows the in situ test on the resistive memory prototype developed in R2RAM from the software point of view, where three different applications run in parallel: 1) tester of the memory, where patterns are provided to memory for a scan of the resistive cells; 2) acquisition of cell currents to make an evaluation not only of the logic values but also of the currents flowing and 3) total dose calculator providing on-line the total dose of the irradiation session.

The test shown in Figure 7.36 used a dose rate of 2 rad/s (Si) with a final total dose of 500 krad (Si) telling that resistive cells based on HfO_2 are practically immune to gamma irradiation [16]. If a suitable RHBD methodology for peripheral parts is followed, the result can be a real rad-hard component able to look to the Mrad (Si) TID.

7.6 Resistive OTPs (ROTPs): The ReRAM Pioneers

The intrinsic immunity to radiations is for sure very appealing, but it cannot be hidden that resistive memories are still in the early stage of their life. Finding the right recipe for manufacturing the MIM stack, the yield problems related to large arrays, the filament growing and dismantling phenomena still not correctly under control show that this technology still must become more and more mature, and if the comparison is with Flash technology, the competition is still lost. Just like all emerging technologies, many silicon runs must be done and many changes in the manufacturing process must be considered, but this is a game lasting not months or years but decades. Flash memories make leverage on a maturity coming from decades of manufacturing development and this has been done when only floating-gate or charge-trap technologies were available. Nowadays, there are many emerging options: phase change, magnetic, resistive and more aggressive approach based on graphene oxide (see Chapter 9). It is only a matter of making them grow and lead to a maturity level sufficient to enter the market.

Despite what is commonly said in research laboratories, the growing process of such technologies it is not only a matter of making a smart manufacturing or a new recipe but also getting the feeling of customers potentially interested in new technologies but needing to make business and satisfying their market. The interaction between research teams and customers is always beneficial for obtaining a homogenous growth of new technologies towards some specific markets. If the issue is to make competition with Flash memories, resistive memories (but also other emerging) have already lost, but stopping the growing process means killing a technology that can be used in different scenarios or that simply requires more time. Unfortunately, many of these new technologies when have demonstrated a proof of concept rising their Technology Readiness Level (TRL) to an intermediate level enter a sort of Death Valley: they are not new and sounding, but they are not mature enough to enter the real market; they are not interesting for scientists looking for new toys to be sold as a Holy Grail, but they are not reliable to become products. Resistive memories, just like others, are in the Death Valley; they still need fuel to boost the process towards maturity, but nobody is keen to take the risk. Why risking in a new technology when we can spend money for gaining something from an old and reliable technology? And the bad news is that in this maturation process costs increase constantly: the higher is the maturity achieved, the higher is the cost. Getting out from this loop is not easy also considering that the gap between Flash and emerging memories risks to increase in favour of the first. But leaving the emerging technologies in the Death Valley can be risky itself simply because all the good things they can offer can be wiped out only because today they cannot replace Flash memories in USB sticks.

Figure 7.37 shows a possible scenario related to resistive memories (but this can be done also for other emerging technologies). Considering the potentialities, not still developed, simply taking into account the physics

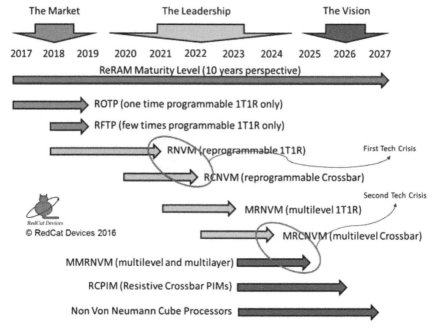

Figure 7.37 Resistive technology roadmap (Copyright RedCat Devices Srl).

behind it, it is not difficult to imagine that crossbars will be the next wave of memory cells built in BEOL as 1T1R; and also the concepts of filament growing can be easily extended to the multilevel approach. Flash memories are already proposing the two-bits-per-cell approach, but in this case, the technology is pushed to the limits and the lower will be the technology node, the lower will be the number of electrons injected in floating-gates or in floating-traps. Resistive memories (but also magnetic and other emerging) have the physics behind and making the multilevel approach can be intrinsically easier, and the concept can be extended on stack of cells. If the recipe is good for making a cell between metal 2 and metal 3, why not between metal 3 and metal 4 and up to metal-N? Nowadays, commercial processes propose up to 10 metal levels and in the future much more: how many bits can be stored in a multi-stack of multilevel resistive cells? But it is not a matter of making storage devices only. Resistive memories can represent the new wave of a generation able to override some bottlenecks of the past like Von Neumann approach to core processors where the continuous passage back and forth of data between logic and storage units (sometimes very far each to the other) limits the performance of such devices. Making closer memories and logic units by using a BEOL technology can be beneficial, at least for the physical reduction of the distance, but it can be beneficial also from the point of view of density. Having the FEOL dedicated to processing and BEOL to storage opens interesting scenarios in the arena of core processors making use of the Process in Memory (PiM) [17] or Computing in Memory (CiM) [18] approach. It is only passing through these progressive steps that one day the Von Neumann approach (belonging to the 1940s) will be overcome leaving space to a new concept of processor having in the FEOL the main computational core unit and in the BEOL a thick stack of storage units able to make themselves parts of computation tasks as sub-core units.

Today, we are only at the beginning of this journey and the probability to see failing what above described is high: market represents insidious and strange waters and new technologies must learn how to swim. But hopefully survival may come also from the market itself. Making leverage on niches (e.g. space market) or on niche applications (e.g. OTPs), a roadmap can be drafted. Today, a complete reprogrammable resistive memory is far to come for the reason described above but a Resistive OTP (ROTP) may be the first attempt to create an interaction with end users. Making use of the differential cell and making leverage, on the one hand, of the unformed cell (having an HRS in the order of $M\Omega$) and, on the other hand, of a formed cell (having an LRS in the order of few $k\Omega$) without any further resetting and/or setting takes the device in the optimal conditions to be only read. Of course, OTP represents a very small market, but it can also be the entry point to gain fuel for the reprogrammable approach; they have embedded the same technology, which in few years can bring to Few Time Programmable (FTP) memories and then to fully reprogrammable memory, to multilevel memories, to crossbar and so on. Making all these steps in laboratory targeting the Holy Grail is impossible: too much effort in terms of costs and, on top of all, no perspective from the market; without long-term vision, there are not possibilities. Better to trust in pioneers and let them be the vanguard of the new wave.

References

[1] A. Grossi et al., "Impact of intercell and intracell varibility on forming and switching parameters in RRAM arrays", IEEE Transactions on Electron Devices, Aug 2015, pages 2502–2059.

[2] K. Kim et al., "Nanofilamentary resistive switchingin binary oxide system; a review on the present status and outlook," Nanotechnology, vol. 22, June 2011.

[3] T. H. Yan et al., "A 130.7 mm2 2-Layer 32 Gb ReRAM Memory Device in 24 nm Technology". In Proceedings of IEEE International Solid-State Circtuit Conference, pages 210–212, Feb. 2013.

[4] Horizon2020 640073 R2RAM Project. Final Report.

[5] Y. Cassuto et al., "Sneak-path constraints in memristor crossbar arrays". In Proceedings of IEEE International Symposium on Information Theory, pages 156–160, 2013.

[6] C. Xuy et al., "Overcoming the challenges of crossbar resistive memory architectures". In *Proceedings of IEEE 21st International Symposium on High Performance Computer Architecture (HPCA'15)*, pages 1–13, 2015.

[7] G. Anelli et al., "Radiation tolerant VLSI circuits in standard deep sub-micron CMOS technologies for the LHC experiments: Practical design aspects". In *IEEE Transactions On Nuclear Science*, 46(6):1690–1696, Dec. 1999.

[8] C. Walczyk et al., "Impact of Temperature on the Resistive Switching Behavior of Embedded HfO$_2$-based RRAM devices", IEEE Trans. Electron Devices, 58, 3124, 2011, pages 3124–3131.

[9] C. Calligaro et al., "An integrated rad-hard test-vehicle for embedded emerging memories", 2016 IEEE International Conference on Electronics, Circuits and Systems (ICECS), pages 5–8.

[10] C. Calligaro et al., "Design of Resistive Non-Volatile Memories for Rad-Hard Applications", IEEE International Symposium on Circuits and Systems, ISCAS, pages 1594–1597, 2016.

[11] F. Wrobel et al., "Simulation of Nucleon-Induced Nuclear Reactions in a Simplified SRAM Structure: Scaling effects on SEU and MBU Cross Sections". In *IEEE Transactions on Nuclear Science*, 48(6), pages 1946–1952, 2001.

[12] A. D. Tipton et al., "Multiple-Bit Upset in 130 nm CMOS Technology". In *IEEE Transactions on Nuclear Science*, 53(6), pages 3259–3264, 2006.

[13] C. Calligaro et al., "Radiation Hardened 2Mbit SRAM in 180nm CMOS Technology". In *Proceedings of IEEE First AESS European Conference on Satellite Telecommunications (ESTEL'12)*. Pages 1–5, 2012.

[14] C. Calligaro et al., "Reading circuit for semiconductor memory cells", United States Patent 5,883,837 March 16, 1999.

[15] A. Parlato, E. A. Tomarchio, C. Calligaro, C. Pace, "The Methodology for Active Testing of Electronic Devices under the Radiations", Nuclear Technology & Radiation Protection: Year 2018, Vol. 33, No. 1, pp. 53–60.

[16] Horizon2020 640073 R2RAM Project: Final Report. January 2017.

[17] G. H. Loh et al., "A Processing-in-Memory Taxonomy and a Case for Studying Fixed-function PIM", Proceedings of WoNDP: 1st Workshop on Near-Data Processing, December 8th, 2013 Davis, California (USA).

[18] S. Jain, A. Ranjan, K. Roy, A. Raghunathan, "Computing in Memory with Spin-Transfer Torque Magnetic RAM", IEEE Transactions on VLSI. Vol. 26 N. 3 March 2018.

8

Technologies for Rad-hard Resistive Memories

Christian Wenger

IHP, Im Technologiepark 25, 15236 Frankfurt/Oder, Germany

> "Science is made up of so many things that appear obvious after
> they are explained."
>
> Frank Herbert, Dune

This chapter presents some technological aspects related to the radiation-hardened design of resistive memories (ReRAM). After an overview of main emerging non-volatile technologies, reliability issues related to a 1T-1R-based HfO_2 (hafnium dioxide) bit-cell are discussed in detail. On an integrated CMOS prototype, several reported experimental data have demonstrated the suitability of such technology for the design of reliable non-volatile devices for radiation environments, such as space.

8.1 Non-Volatile Memory Technologies

Present memory technologies, including DRAM, SRAM and Flash, are potentially approaching their scalability limits. Accordingly, the ITRS (International Technology Roadmap for Semiconductors) has recently completed an assessment of eight memory technologies among the emerging devices and recommended that Resistive RAM (ReRAM) receives additional focus in research and development. ReRAM has shown promising scalability, non-volatility, multiple state operation, 3D stackability and CMOS compatibility.

Furthermore, for embedded applications, it is mandatory to achieve low-power dissipation, short access time, bit alterability, high reliability at room and elevated temperatures and resistivity towards radiation. However, there are still some challenges for ReRAM devices, especially long-term reliability and a better understanding of the microscopic picture of the switching. The reliability issue includes switching endurance and thermal stability. The challenges can be tackled through in-depth investigation of highly scaled memory element designs as well as through identification of the major sources of process variations.

On the technology and device level, the utmost objective is to improve memory element structure, size and selection device for ReRAM in terms of radiation hardness. This includes the trade-offs within the MIM structure geometries, scaling of the memory element size usually leading to increased R_{OFF}/R_{ON} ratio and improved long-term reliability. However, even more important is the integration of the ReRAM in a large array with the use of selection devices to isolate the ReRAM elements one from another. The established method is to use a MOSFET in series with the ReRAM element, in a so-called 1T-1R arrangement. The current supplied by the selection device must exceed the required reset current of the memory element. Unfortunately, MOSFETs have limited current drive capability. One way to solve this problem is to simply make the access device larger driving a larger current. However, this reduces memory density. Thus, to ensure sufficient reset currents, the trade-off analysis between minimal-width MOSFETs and highly scaled ReRAM elements is necessary.

Different types of variability can affect the operation, performance, endurance and reliability of ReRAM devices, ranging from inter-cell variability introduced during processing, inter-cell variability as resistances change over time after programming and intra-cell variability – cycle-to-cycle variation of the set and reset resistances of any given cell. While the read voltage, reset and set pulse can be optimized for the memory cell, variations between cells must be minimized. Process-induced variability in the physical structure of the ReRAM element can result in devices which react differently to the same stimuli. The structural and physical properties, thickness and uniformity of metal/insulator layers play significant role. All reliability and variability effects should be investigated on an array structure for the sake of actual manufacturability and practical viability of this memory technology.

8.1.1 State-of-the-Art NVMs – Flash Memory

Flash memories have been widely adopted in various consumer and even enterprise products in recent years because of its non-volatile, shock-resistant and power-economic characteristics. A flash memory is an NVM built from charge storage elements called floating gates. Writing and erasing is performed by injection of a charge (writing logic 0) into and its removal (writing logic 1) from the isolated floating gate. Both injection and removal of a charge require high voltage (HV) to be applied. Because HV has negative impact on the floating gate, programming operations are destructive over time, decreasing memory's reliability. With the density advancement of semiconductors, high-density and low-price Multi-Level Cell (MLC) flash chips have replaced the role of Single-Level Cell (SLC) flash chips in many applications. However, compared to SLC flash memory, MLC flash memory has lower read/write performance, higher bit error rate and lower endurance.

Much research has been conducted concerning the reliability of flash technology. It is known that aggressive scaling reduces the number of electrons stored in a floating gate, which causes retention problems [1]. Scaling flash below 20 nm is projected to be very difficult [2]. Decreasing the cell size causes coupling between adjacent cells and quantum effects, which can change the write characteristics and deteriorate data integrity [3]. Reliability of floating gate devices is also a complex problem. There are many fault mechanisms present in Flashes, but to simplify modelling of the flash memory, only disturbance errors are taken into consideration, since they are the most frequent errors [4]. Disturbance failures, which are not destructive, affect neighbour pages or data in a page where memory operations are performed. The typical endurance of flash memory is also a significant issue, since it is for MLC flash memory 10^4 programming operations in comparison to 10^5 for the SLC. In addition, MLC devices have longer and variable programming latencies; consume more energy for read/write and more idle power. Besides that, flash memories are susceptible to complex defects and process variations and are prone to radiation-induced failures. Moreover, the performance of the Flashes further restricts the Radiation hard use of the devices due their very asymmetric behaviour with very long writing cycles.

Flash is currently without any doubts the leading NVM technology in both consumer and embedded markets. However, the problems imposed with the scaling, performance limits and endurance issues accelerated the development of the emerging novel NVMs, which should on a long term not only replace Flash memories as main NVM driver but also completely change classical architectural paradigm of NVM/RAM combination on the chip. As a result, the emerging technology will provide for a variety of applications the foundation for building the systems based on the Radiation hard memory approach.

8.1.2 The Way-out Over Emerging Technologies

Several non-volatile memory device concepts have been proposed in the last few years. In ferroelectric RAM (FeRAM) cells [5], data are stored by the bistable switching behaviour of the polarization of a ferroelectric layer under external voltage pulses. Despite the promising characteristics, FeRAM suffers, in particular, from compatibility problems with mainstream Si technologies, namely etching and contamination issues are difficult to control due to the complicated stoichiometry of ferroelectric materials [6]. Another concept is provided by magnetoresistive RAM (MRAM) cells [7], which are based on the magnetoresistive effects in magnetic materials and structures that exhibit a resistance change when an external magnetic field is applied. However, it has been pointed out [8, 9] that the biggest issue for the integration of this memory technology in future products is mainly related to the low geometrical shrinking capabilities of the memory, thus reducing the MRAM scalability. Phase-Change Memories (PCMs) are relying on the phase transition from crystalline to amorphous and vice versa of a Germanium-Antimony-Tellurium (GeSbTe or GST) chalcogenide material [10]. Probably the biggest challenge for phase change memory is its long-term resistance and threshold voltage drift [11]. The resistance of the amorphous state slowly increases according to a power law ($\sim$t0.1). This limits the ability for multilevel operation and could also jeopardize standard two-state operation if the threshold voltage increases beyond the design value.

Finally, in R^2RAM (Radiation Hard Resistive Random-Access Memory) project [12], funded by the European Union's Horizon2020 Programme, we focused on ReRAM technology, which will be evaluated in detail in the next section. This technology is already seen as a major candidate for Radiation-Hard memory devices [13]. In Table 8.1, the emerging NVM technologies are benchmarked with their expected development.

Table 8.1 Benchmark of emerging NVM technologies

Device Type	SRAM	DRAM	NAND Flash	HDD	STTRAM	PCRAM	ReRAM
Maturity	Product	Product	Product	Product	Prototype	Prototype	Prototype
Cell size	$150F^2$	$6F^2$	$4F^2$	$(2/3)F^2$	$6F^2$	$4F^2$	$4F^2$
MLC capability	No	No	4 bits/cell	No	2 bits/cell	4 bits/cell	4 bits/cell
Write energy	$1pJ$	$2pJ$	$10nJ$	N/A	$0.02pJ$	$100pJ$	$0.1pJ$
Write latency	$5ns$	$10ns$	$200\mu s$	$10ms$	$20ns$	$100ns$	$20ns$
Write endurance	10^{16}	10^{16}	10^5	N/A	10^{16}	10^8	10^8

8.1.3 ReRAM Technology

The ultimate concept for the Radiation-Hard memory resides on the ReRAM devices. The resistive RAM concept is based on a non-volatile memory phenomenon realized by reversibly changing the resistance of a thin film layer material sandwiched between two metal electrodes by applying external voltage pulses. The Ox-ReRAM NVM approach is a special case in which the resistive switching of oxide thin film structures in Metal-Insulator-Metal (MIM) cells is exploited. ReRAM are the candidates in replacing Flash devices due to their cost effectiveness (Back-End-Of-Line, BEOL, integration allows to save valuable wafer area), fastest read/write times, low power and high reliability.

Active ReRAM arrays, in which each cell contains 1T-1R, have already been demonstrated by various research groups. In IHP's 1T-1R configuration, the drain of the transistor is connected to the bottom metal electrode of the MIM cell at each storage location (Figure 8.1(a)). The cell is selected by turning on the transistor by applying a voltage to the appropriate row and simultaneously biasing the corresponding column. The Set operation changes the ReRAM device from a high-resistive state (OFF-state) to a low-resistive state (ON-state) by applying a voltage to the bit-line (BL) and 0V to the source terminal of the transistor. The Reset operation changes the device to OFF-state by applying a voltage to the source terminal of the transistor and 0V to BL.

The most advanced ReRAM integration has been recently demonstrated by the ITRI (Industrial Technology Research Institute) at ISSCC 2011 [14].

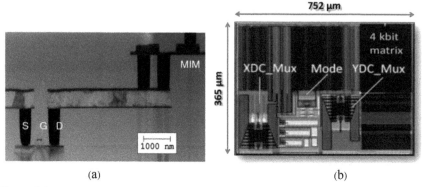

Figure 8.1   TEM (Transmission Electron Microscopy) cross section with MIM cell with transistor (a) and 4 kbit ReRAM architecture (b).

Based on $TiN/HfO_2/Ti/TiN$ ReRAM structure, they have demonstrated the entire memory chip based on 32×128 kbit sub-blocks with the associated control circuitry. An integration density of 4 Mbit as well as switching times to sub-nanosecond level, robust cycling endurance in the 10^{12} range, high device yield and an extrapolated data retention higher than 10 years at $220°C$ have been shown. Both operation voltages (V_{Set} and V_{Reset}) are reported to be less than 2 V, and the resistance ratio between OFF-state and ON-state is larger than 20 [15]. Additionally, the HfO_2-based ReRAM devices allow for the possibility of multi-level operation, i.e. the use of more than two states in one cell.

IHP [16] was among the first groups to establish resistive switching in HfO_2-based ReRAM structures [17]. Recently, MIM memory devices in a 1T-1R architecture have been integrated in a modified Si CMOS technology platform, as shown in Figure 8.1(a) [18, 19]. Reliable resistive switching characteristics were achieved in this 1T-1R architecture with fast switching times < 100 ns, a resistance ratio R_{OFF}/R_{ON} of more than 10 and temperature stability up to $125°C$. To evaluate the potential of ReRAM, a 4 kbit memory device array with periphery was specified and designed (Figure 8.1(b)). The results achieved until now show that ReRAM can provide low area occupation, MLC capability, low write energy, low write latency and a write endurance characteristic that combines the advantages of Flash and DRAM products – pushing this technology in the front line for Radiation-Hard memory.

What is still an open issue being the trade-off analysis between minimal-width select devices and highly scaled ReRAM elements. The area of an individual 1T–1R cell is defined by the select transistor area, which is typically $15F^2$–$20F^2$ (where F is the minimum feature size). The cell size has a lower limit of 6–8 F^2 corresponding to the lateral placement of the three terminals side by side plus the required isolation.

The main limiting factor to ReRAM array integration and to ramp-up in product maturity level is ascribed by the multiple ways to implement an ReRAM memory cells. Most of the array investigations for ReRAM have been performed on large test structures – arrays without control logic for addressing and write/erase the memory, whereas very few solutions with fully addressing capabilities have been proposed. The need to improve non-volatile memories reliability in embedded systems is an additional design concern. The reliability-performance trade-off is obtained by partitioning the memory addressable space in different functional blocks, each written by means of a specific optimized writing algorithm. In particular, it has been demonstrated in Ref. [20] that multiple ways exist to achieve the erasing operation in resistive sensing memories. This represents a degree of freedom, which can be exposed to the upper layers of the design hierarchy to implement technology-based selective data protection and therefore exploit reliability/speed trade-off, which has not been demonstrated so far. However, the electrical properties of integrated 1T-1R devices as well as those of a memory array significantly differ. This is expressed in uncontrollable random parametric variations from device to device and from chip to chip. Hence, device uniformity of ReRAM structures must be further improved to meet the manufacturing requirements for emerging Radiation hard memory technologies.

8.2 Reliability Issues of 1T-1R-based HfO₂ ReRAM Arrays

A comparison between 1T-1R ReRAM 4kbits arrays manufactured using either amorphous or polycrystalline HfO_2 in terms of performance, reliability, Set/Reset operations energy requirements and intra-cell and inter-cell variability during 10k Set/Reset cycles has been performed. The polycrystalline film resistor area is equal to 0.4 μm^2. For amorphous films, also a resistor with larger area has been considered, which shows higher reliability and performance (i.e. 1 μm^2). The Forming/Set/Reset operations on the arrays were performed by using the Incremental Pulse and Verify algorithm.

1T-1R cell arrays integrated with Amorphous-HfO_2 (A-array) with small (0.4 μm^2) and large (1 μm^2) resistor area and Poly-crystalline-HfO^2 (P-array) resulted in a Forming Yield (calculated as the cell percentage having

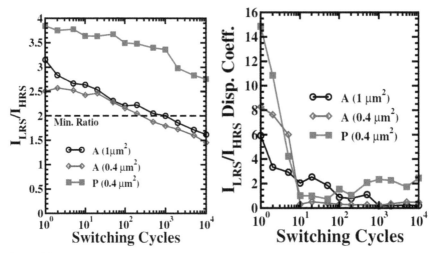

Figure 8.2 I_{LRS}/I_{HRS} current ratio average values (left) and dispersion coefficients (right) calculated during cycling.

a read verify current after forming $I_{read} \geq 20\mu A$) of 58%, 90% and 95%, respectively. Figure 8.2 shows the average current ratios between Low Resistive State (LRS) and High Resistive State (HRS) read currents (I_{LRS}/I_{HRS}), calculated on the entire cells population during Set/Reset cycling, and their relative dispersion coefficient. The dispersion coefficient, defined as (σ^2/μ), has been used to evaluate the cell-to-cell variability. The minimum current ratio that allows to correctly discriminate between HRS and LRS ($I_{LRS}/I_{HRS} > 2$) is indicated for comparison purposes. Due to the faster cell degradation, the average ratios of A-arrays with resistor areas of 0.4 and 1 μm^2 cross the minimum ratio limit after 200 and 1 k cycles, respectively. P-array showed higher ratio (≈ 2.8) even after 10 k cycles, but also a higher dispersion coefficient after forming (i.e. cycle 1). The grain boundaries conduction mechanism in the polycrystalline HfO$_2$ structure could be the reason of the higher cell-to-cell variability in P-arrays. A-array with resistor area of 1 μm^2 shows a slightly higher average ratio and a slower degradation than A-array with resistor area of 0.4 μm^2. In smaller cells, the presence of defects in the HfO$_2$ stack has a stronger impact on the performance since it makes the switching operations more difficult to control, speeds up the degradation and increases the overall inter-cell variability.

Figure 8.3 shows a comparison between I_{LRS} and I_{HRS} cumulative distributions measured at cycle 1 and after the 10 k Set/Reset cycling test: A-arrays show more compact distributions at cycle 1; however, after cycling,

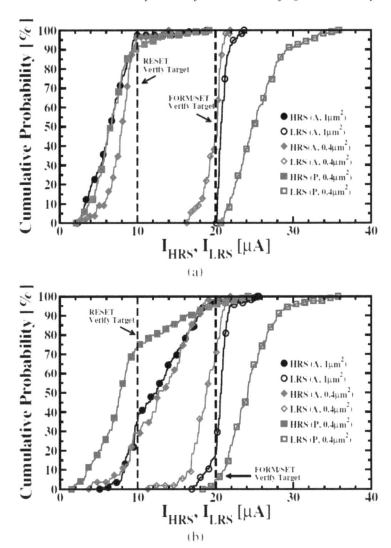

Figure 8.3 I_{HRS} and I_{LRS} cumulative distributions at cycle 1 (top) and at cycle 10 k (bottom).

P-array shows a higher percentage of correctly switching cells reaching the Set/Reset verify targets. I_{HRS} cumulative distribution in P-array shows a larger distribution tail at cycle 1 compared to A-arrays. After 10 k cycles, the cells degradation makes it more difficult to break or re-create the filament; hence, the voltage requested to reach the verify target increases as well as the number of cells not able to reach the verify target. An enlargement of the upper tail in P-array HRS distribution can be observed, whereas on A-arrays, a strong shift of the distributions towards higher currents occurs, since a higher number of cells is not able to reach the Reset threshold. The reason of the lower ratio in A-array with small resistor area can be explained by the cumulative distributions, since they show lower I_{LRS} and higher I_{HRS} than cells with larger resistor area either at cycle 1 or after 10 k cycles. In I_{LRS} cumulative distributions, the cells not able to reach the Set verify target generate a lower tail on P-arrays after 10 k cycles, whereas on A-arrays, a higher number of cells is not able to reach the Set verify target especially when cells with resistor area of 0.4 μm^2 are considered. These results in a strong shift of the distributions towards lower currents, especially in A-array with small resistor area that shows a large number of cells not reaching the Set verify target even at cycle 1.

Figure 8.4 shows the average Set and Reset switching voltages (V_{SET}, V_{RES}) and their relative dispersion coefficients: lower V_{SET} and V_{RES} are required on P-array, which also shows no variations during Set/Reset cycling, whereas V_{SET}, V_{RES} increase on A-arrays during cycling. V_{RES} on P-array shows the highest variability: such operation is critical and very difficult to control in ReRAM arrays since it strongly depends on how the filament is created: over-forming, as well as endurance degradation, can make the filament difficult to disrupt, increasing the V_{RES} variability. A-arrays show similar behaviour of the average V_{SET} and V_{RES} (a lower average V_{SET} is observed on A-array with larger resistor area only up to 500 cycles) while a higher V_{SET} and V_{RES} dispersion can be observed in A-array with smaller resistor area.

Figure 8.5 shows the cumulative distributions of switching voltages, V_{FORM}, during Forming, while Figure 8.6 shows the Set and Reset switching voltages cumulative distributions at cycle 1 and after the Set/Reset cycling. Forming, Set and Reset incremental pulse algorithms starting point and last attempt are indicated, corresponding to the first and the last voltage pulses available in the incremental Pulse and Verify procedure.

P-array requires lower V_{SET} and V_{RES} but higher V_{FORM} if compared to A-array with the same resistor area. A-array with larger resistor area requires higher V_{FORM}. Moreover, it can be observed that $\approx 40\%$ of the devices with smaller resistor area reached the forming threshold at $V_{FORM} = 2V$,

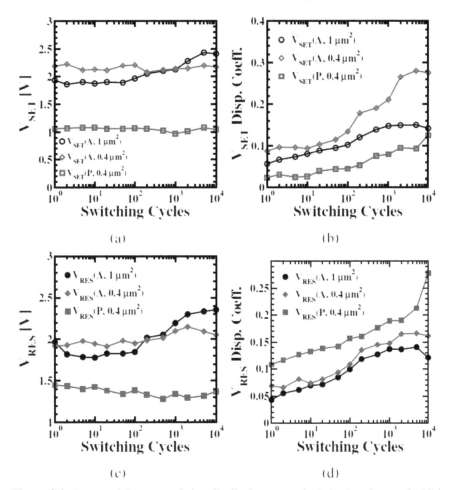

Figure 8.4 I_{HRS} and I_{LRS} cumulative distributions at cycle 1 (top) and at cycle 10 k (bottom).

corresponding to the first attempt of the Forming Algorithm. Since P-array shows a more compact distribution on V_{SET} and a larger V_{RES} than A-arrays, faster Set operation could be reliably used on P-array, whereas on Reset, an incremental pulse with verify technique is required to ensure good reliability. A-arrays show large distributions on both V_{SET} and V_{RES}, hence the adaptation of incremental pulse with verify techniques is mandatory on such arrays.

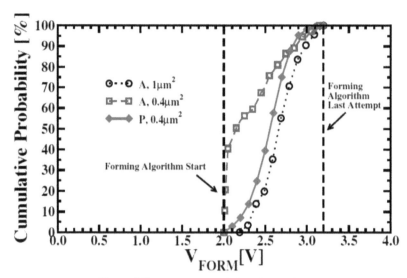

Figure 8.5 V_{FORM} cumulative distributions.

Figure 8.7 shows the average energy required to perform Set and Reset operations on a single cell. The overall energy required to create/disrupt the conductive filament during Set/Reset operations has been calculated as:

$$E = \sum_{i=1}^{n} V_{pulse,i} \times I_{pulse,i} \times T_{pulse} + V_{read} \times I_{read,i} \times T_{read} \qquad (8.1)$$

where n is the number of set/reset pulses applied during incremental pulse operation, $V_{pulse,i}$ is the pulse voltage applied at step "I", $I_{pulse,i}$ is the current flowing through ReRAM cell during pulse "I" application, $T_{pulse} = 10 \ \mu s$ is the pulse length, $V_{read} = 0.2$ V is the read voltage applied during verify operation, $I_{read,i}$ is the current read on the ReRAM during read verify step "I" and $T_{read} = 10 \ \mu s$ is the verify pulse length.

P-array shows lower power consumption with a lower increase during cycling thanks to a lower V_{SET} and V_{RES}. A-arrays with different resistor areas show similar power consumption during Reset operation, whereas a lower consumption during Set is observed on A-array with larger resistor area only up to 500 cycles since cells with larger resistor area require lower V_{SET}.

In conclusion, 1T-1R ReRAM arrays manufactured with P-HfO$_2$ shows several advantages compared to A-HfO$_2$ even considering their improved process: higher current ratio, lower switching voltages, lower power

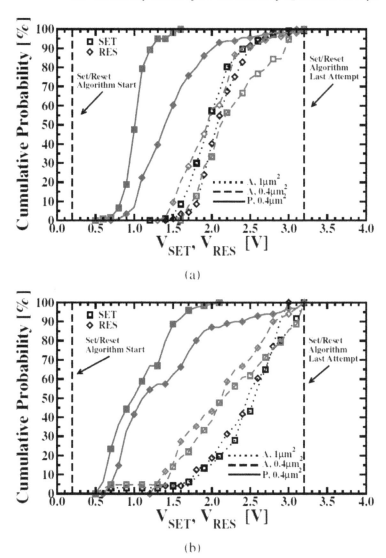

Figure 8.6 V_{SET} and V_{RES} cumulative distributions at cycle 1 (top) and at cycle 10 k (bottom).

consumption, minor endurance degradation and higher overall yield. Moreover, P-array shows very low V_{SET} variability; hence, faster Set operation could be reliably performed. P-array disadvantages are represented by the larger HRS distribution after Forming, the higher Reset voltage dispersion and the higher V_{FORM} if compared to A-array with the same resistor area;

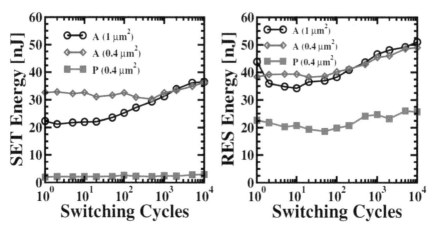

Figure 8.7   Energy required to perform Set (left) and Reset (right) operations as a function of the Set/Reset cycle number.

however, it must be pointed out that Forming operation is performed only once. The grain boundaries conduction mechanism in the polycrystalline HfO$_2$ structure could be the reason for the higher cell-to-cell variability in P-arrays.

8.2.1 Retention Results on Amorphous and Polycrystalline Arrays

We investigated the evolution of HRS and LRS distribution obtained through incremental Pulse and Verify algorithm during 100 h retention measurements at 125°C. Both Amorphous (A-, with MIM resistor area of 1 μm^2) and Polycrystalline (P-, with MIM resistor area of 0.4 μm^2) arrays are used. A full characterization of the intermediate cycling condition in retention is performed on polycrystalline, while on amorphous, the retention test was performed only on fresh (at cycle 1) and after 10 k cycles.

8.2.2 Set Evolution

The LRS distribution during retention shows an overall shift and a tail creation in the direction of HRS: this is related to the relaxation of the conductive filament. In extreme cases, this relaxation causes the rupture of the filament. In this case, the read current should be distributed around

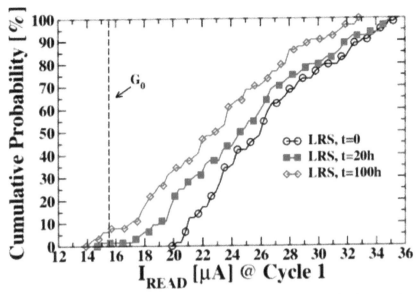

Figure 8.8 Set cumulative distributions evolution in retention after 1 Set/Reset cycle.

the quantum conductance unit G$_0$, which is the limit between the presence/absence of the conductive filament and hence the point of maximum instability. As an example, Figure 8.8 shows the evolution during retention of LRS cumulative distributions obtained on Polycrystalline array after 1 Set/Reset cycle.

A similar behaviour is observed after cycling, as shown in Figure 8.9. The left tail in the distribution increases with cycling, resulting in an initial decrease of the average LRS current and an initial increase of the dispersion coefficient. However, the general trend in retention is the same for both average value and dispersion coefficient in each cycling condition.

A comparison between LRS retention behaviour in Amorphous (A) and Polycrystalline (P) samples in terms of average read current ad dispersion coefficient is reported in Figure 8.10. Lower read currents and dispersion coefficients are observed on the Amorphous sample; however, the trend observed during retention is the same that of Polycrystalline for both average and dispersion, confirming the validity of the analysis also for different resistor areas and HfO$_2$ stacks.

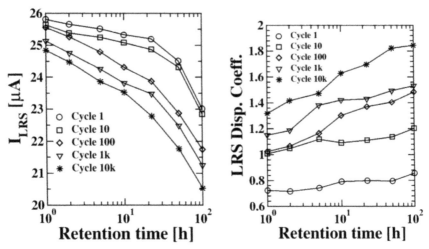

Figure 8.9   LRS read current average value (left) and dispersion coefficient (right) evolution during retention, in different cycling condition.

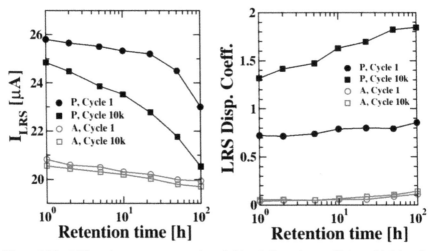

Figure 8.10   LRS read current average values (left) and dispersion coefficients (right) evolution during retention after 1 and 10 k Set/Reset cycles for both Amorphous and Polycrystalline samples.

8.2.3 Reset Evolution

The HRS distribution during retention shows an overall shift (indicated as mean shift) in the direction of G_0. However, the behaviour in retention is the opposite on the cells that failed to Reset correctly (with an initial read

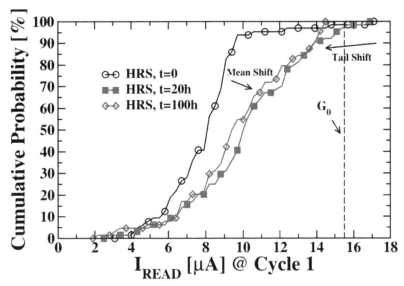

Figure 8.11   Reset cumulative distributions evolution in retention after 1 Set/Reset cycle.

current higher than G_0): since the presence of a conductive filament can be assumed on these cells, the behaviour during retention is the same as previously observed in Set and causes a decrease of the current (indicated as tail shift). As an example, Figure 8.11 shows the evolution during retention of HRS cumulative distribution obtained on the Polycrystalline array after 1 Set/Reset cycle.

With the increase of the Set/Reset cycles (see Figure 8.12), the number of cells that fail to Reset increase (as well as the tail shift contribution in retention), while the number of cells that correctly reset decrease (as well as the mean shift contribution). This explains why the trend of HRS dispersion coefficient changes with cycling: on fresh devices, only a few cells fail to Reset; hence, the initial dispersion is low as well as the average HRS current and the overall retention effect is the mean shift, resulting in an increase of both average and dispersion coefficient. When the number of cycles increases, more cells fail to Reset; hence, the initial average HRS current and dispersion increase and the Tail Shift contribution becomes dominant, causing a reduction of the dispersion during retention and compensating the average HRS read current increase. Again, the same considerations can be applied on the amorphous sample, as shown in Figure 8.13.

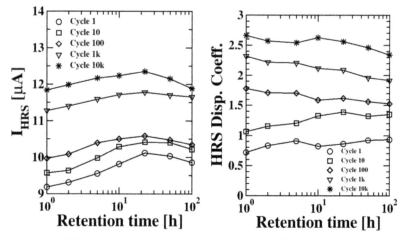

Figure 8.12 HRS read current average values (left) and dispersion coefficients (right) evolution during retention, in different cycling conditions.

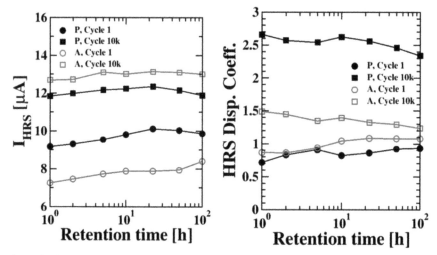

Figure 8.13 HRS read current average values (left) and dispersion coefficients (right) evolution during retention after 1 and 10 k Set/Reset cycles for both Amorphous and Polycrystalline samples.

In conclusion, we can state that both LRS and HRS tend to collapse around G_0 during retention for both HfO_2 stacks and MIM resistor areas. Cycling seems to impact only the initial condition of the distributions before the retention test, while the retention behaviour could be explained in the same way in each cycling condition.

8.2.4 Impact of Temperature on Conduction Mechanisms and Switching Parameters in HfO₂-based 1T-1R ReRAM Devices

The impact of temperature in the range of $-40°C$ to $+150°C$ on the leakage mechanism and resistive switching voltages of 1T-1R HfO₂-based devices is investigated. For this purpose, a complete characterization in that temperature regime was performed in AC mode through the Incremental Step Pulse with Verify Algorithm (ISPVA). In addition, DC mode characterization was performed in the same temperature range, to extract the most relevant Quantum-Point Contact (QPC) model parameters such as tunnel barrier height and width from the Current-Voltage curves.

To show the impact of the temperature on the inter-cell variability, Figure 8.14 illustrates the cumulative probability of the measured currents in the AC mode by applying the read and verifies operation. As shown in this figure, the HRS currents are not impacted by temperature. In contrast to the HRS state, there is a clear increase of the dispersion and the mean value of the LRS current state with decreasing temperature.

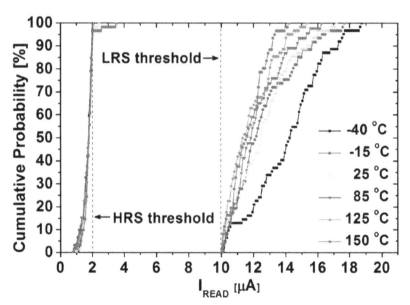

Figure 8.14 HRS and LRS current cumulative probabilities in AC mode at the selected temperature values. The target current values for each operation are also marked.

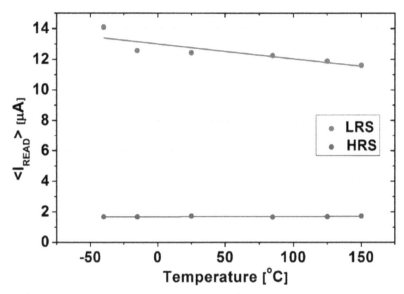

Figure 8.15   Average values of the HRS and LRS currents and their dependence with temperature variations.

The average current values of the HRS and LRS as a function of temperature are illustrated in Figure 8.15. Concerning the LRS, a slope of about -10^{-2} μA/°C was obtained, which means that the conduction mechanism of the LRS is metallic-like. The slope of the HRS is, with 10^{-4} μA/°C, 100 times lower than the LRS one, suggesting a tunnel-like conduction mechanism.

As Figure 8.16 illustrates, there is no significant trend with temperature in switching voltage values measured in AC mode, neither in Reset (V_{Res}) nor in Set (V_{Set}) operations. Therefore, a temperature independence of the switching voltage is established. This temperature independent behaviour of both operations provides very good device stability in the relevant temperature range, which is desirable for applications of these devices.

The measurement mode only affects the filament conduction in a quantitative way but not in the conduction mechanism itself. Therefore, Current-Voltage HRS characteristics measured in DC mode were used to analyze the conductive filament properties through the QPC model at the considered temperature range. The three main parameters involved in the QPC model are Φ (the barrier height at the constriction), α (related to the barrier curvature, assuming a parabolic longitudinal potential, thus with the barrier width) and β (which considers how the potential drops at the two ends of the filament),

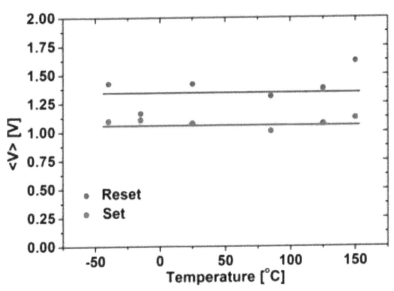

Figure 8.16 Average values of the Set and Reset switching voltages and their variation with temperature.

as can be seen in the following equation that represents the model of HRS current:

$$I_{HRS} = \frac{2e}{h}\left(eV + \frac{1}{\alpha}\ln\left[\frac{1 + e^{\alpha(\phi - \beta eV)}}{1 + e^{\alpha(\phi + [1 - \beta]eV)}}\right]\right) \qquad (8.2)$$

Figure 8.17 shows Φ and α average values at each temperature. On the one hand, Φ increases when the temperature increases, with a rate of about 2×10^{-3} eV/°C. Therefore, a current decrease is expected at higher temperatures. On the other hand, α decreases when the temperature increases at a rate of about 9×10^{-4} (eV°C)⁻¹. Therefore, a current increase is expected at higher temperatures, because this barrier width decreases.

QPC parameters variation shows two opposite temperature effects that change the barrier shape in the constriction point of the filament in HRS. As it can be seen in Figure 8.18, at higher temperatures, the barrier is higher but narrower, whereas at lower temperatures, the barrier is lower but wider. In LRS, the top of the barrier is supposed to be under the energy reference level (E = 0 eV), the conduction in the constriction point being ballistic-like. This theoretical explanation would be consistent with the fact that the conduction through the oxide in LRS state is dominated by the metallic-like conduction attributed to the filament.

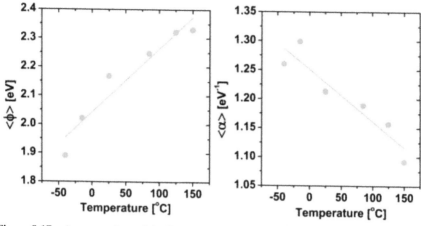

Figure 8.17   Average values of the Φ and α parameters and their variation with temperature.

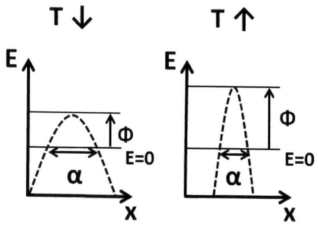

Figure 8.18   Schematic of the change of shape of the barrier at the constriction considering low and high temperatures.

8.3 CMOS Integration of Resistive Memory Cells

The integrated 1T-1R memory bit-cells are constituted by a selector ELT transistor plus the resistive device and are manufactured in IHP 250 nm CMOS technology. The n-type MOS selector whose drain is in series to a Metal-Insulator-Metal (MIM) stack also sets the current compliance. The MIM cell integrated on the metal level 2 of the CMOS process is a TiN/HfO$_2$/Ti/TiN stack of 150 nm TiN layers deposited by magnetron sputtering, a 7 nm Ti layer and a 9 nm HfO_2 layer grown by Chemical Vapor Deposition (CVD). The MIM cells area is 700×700 nm^2. The TEM cross view of the Rad Hard designed 1T-1R cell and the circuit scheme are illustrated in Figure 8.19.

The 1Mbit device was the result of the ensemble of eight 128 kbit modules each having its own decoding scheme, ATD (Address Transition Detection) and sense amplifier. This architectural approach protects the test vehicle against Multiple Bit Upset (MBU) reducing every contribute from charged particle to at least only one Single Event Upset (SEU) as already demonstrated in previous SRAM devices. Even if complete of all decoding schemes, the 1 Mbit (2 Mcells) device also contains Direct Memory Access (DMA) to provide the possibility to characterize the behaviour of ReRAM cells independently from the sensing scheme and setting/resetting circuitry. This freedom degree enables the access via ATE equipment for all analysis related to set and reset state distributions and a better approach to the real characterization of the resistive state of the cell under very different conditions (voltage gate, drain and source).

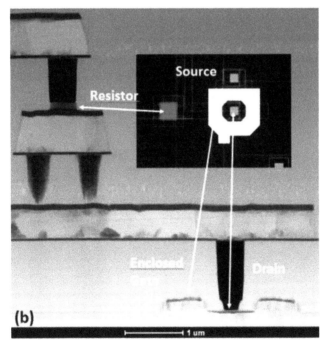

Figure 8.19 TEM cross-sectional image of 1T-1R architecture with an ELT transistor and a 0.7×0.7 μm^2 MIM cell.

To achieve an increased resistance against radiation, the single bit is the result of the contribution of two ReRAM cell located in different array locations; this guarantees an internal redundancy (no reference cells are required for read mode operations) and a wider margin window in a sensing module very similar to SRAMs. The architectural approach is based on where two arrays (left and right) contain left cells and rights cells: logic "1" is the result of a low resistance in the left cell and a high resistance in the right cell and vice versa for logic "0". The differential cell approach, thanks also to the independence of row decoding final stages and column decoding switches, guarantees resistance against Single Event Effects (SEEs) and disturbs in all conditions (Read, Set and Reset modes).

The design of the 1 Mbit chip required an area of about 60 mm^2, as illustrated in Figure 8.20. In total, six wafers were processed with almost 400 process steps and process control steps. To evaluate the impact of radiation, one wafer was diced and packaged in DIL 48. The picture of the assembled 1 Mbit chip is shown in Figure 8.21.

The 1T-1R structure of the memory array consists of n-type MOS access transistors, which are sensitive to radiation. In advanced CMOS technologies with thin gate oxides, the leakage paths beneath the shallow trench corners become the major contributor to the total ionizing dose (TID) effect of NMOS. A suitable approach to eliminate the leakage path in n-type MOS transistors is to adopt a gate-enclosed layout. Therefore, we present a 1T-1R device based on the combination of an Enclosed Layout Transistor (ELT) and a $TiN/HfO_2/Ti/TiN$ based resistor.

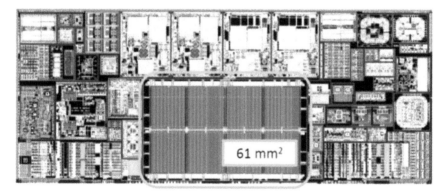

Figure 8.20 Top view of the complete test field including the 1 Mbit memory chip, marked in dark yellow.

Figure 8.21   Picture of the demonstrator assembled into a DIL 48 package (without lid).

To study the electrical performance of such devices, a complete charac-terization was performed in AC mode, through the Incremental Step Pulse with Verify Algorithm, and in DC mode, providing data for the filament constriction modelling using the Quantum-Point Contact model. The drain current versus gate voltage characteristics after irradiation of the ELT is illustrated in Figure 8.22. The leakages are almost unaffected during the total ionizing dose irradiation up to 750 krad.

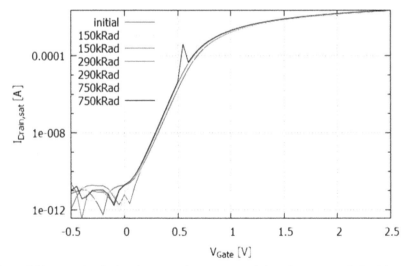

Figure 8.22   Post-irradiation Current Voltage characteristics of the access ELT with varied dose rate.

In the DC measurement mode, double voltage sweeps were applied on the Source/Drain terminals, corresponding to reset/set and forming operation, while recording the I–V curves. The first sweep starts at 0 V and stops at 5 V with step heights of 0.05V, whereas the second one follows the opposite voltage trail recording the current value at 0.2 V defined as HRS and LRS current, respectively. The applied transistor gate voltage values were VG = 2.8 V and VG = 1.3 V, respectively.

In the AC characterization mode, the ISPVA was applied. This technique consists of a sequence of increasing voltage pulses (V_{PULSE} = 0.2–5 V, V_{STEP} = 0.1 V, t_{PULSE} = 10 μs, $t_{FALL/RISE}$ = 1 μs) on the drain terminal during set and forming operation, whereas this sequence of pulses is applied on the source terminal during the reset operation. The applied voltage values V_G (gate) were the same as in DC mode, respectively. After every pulse, a Read-verify operation is performed with V_G = 1.3 V, V_{READ} = 0.2 V (applied at the drain contact) for 10 μs. When the read current reaches the target value of 6 μA, the set and forming operations are stopped, whereas the reset operation is stopped when the target value of 3 μA is achieved.

From the reset I–V characteristics measured in the DC mode, the analysis of the conductive filament properties in the 1T-1R bit-cell using the QPC model was performed. The model is based on the idea that in HRS a constriction point (rupture point) takes place in the filament (Figure 8.23(b)), defined as an energy barrier (Figure 8.23). Thereby, the HRS current can be calculated by the following expression:

$$I_{HRS} = \frac{2e}{h}G/G_0 \left(eV + \frac{1}{\alpha}\ln\left[\frac{1 + e^{\alpha(\phi - \beta eV)}}{1 + e^{\alpha(\phi + [1-\beta]eV)}}\right]\right)$$

where ϕ is the barrier height (bottom of the first quantized level), α is a parameter related to the inverse of the potential barrier curvature (assuming a parabolic longitudinal potential), β defines the position of the constriction point with respect to the two electrodes, G/G_0 is a conductance parameter equivalent to the number of filaments and $G_0 = 2e^2/h$ is the quantum conductance unit corresponding to the creation of a single mode nanowire, where "e" is the electron charge and "h" is Planck's constant.

To activate the resistive switching behaviour, the ReRAM devices require a preliminary forming step. This initial operation plays a fundamental role in determining the subsequent devices performance. In Figure 8.23, the forming voltages (a) and LRS current values (b) achieved in DC and AC modes in 80 devices are compared. There are no substantial differences in the forming

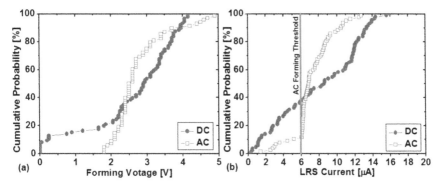

Figure 8.23 Cumulative distributions of forming voltages (a) and LRS currents (b) in both modes.

voltage distributions; nevertheless, the variability of the forming currents is strongly reduced by ISPVA.

The distributions of the reset/set parameters are illustrated in Figure 8.24. In agreement with previous publications, the reset voltages are larger than the voltages required for setting the filaments. Moreover, caused by the lower electrical power applied to the sample in the AC mode, reset/set switching takes place at higher voltages. As shown in Figure 8.24(b), there is no clear window between HRS and LRS in the DC mode. In contrast, the ISPVA lead to an evident gap defined by the current thresholds specified by 3 and 6 μA, respectively. Therefore, the use of such algorithms is mandatory to define the HRS and LRS levels with strongly reduced cell-to-cell variability.

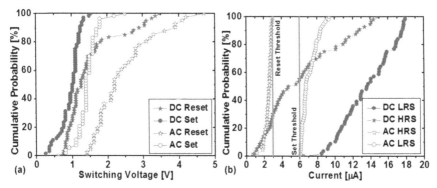

Figure 8.24 Cumulative distributions of the reset/set voltages (a) and HRS/LRS currents (b) in DC and AC modes.

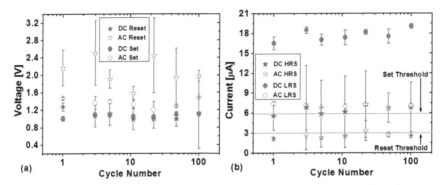

Figure 8.25   Average and dispersion values of reset/set voltages (a) and of HRS/LRS currents (b), received by applying DC and AC modes as a function of cycling.

The endurance performance of the best 11 devices has been verified by 100 cycles. Good switching voltage stability is shown in Figure 8.25(a), except for reset in the AC mode. During cycling, the current values remain stable. According to Figure 8.25(b), the DC mode provides better current ratio ($\sim$3) than the AC mode ($\sim$2). Nevertheless, the average values shown in Figure 8.24(b) do not consider the higher cell-to-cell variability in the DC mode. Therefore, the ISPVA remains as a better approach to program the devices.

To study the evolution of the filament constriction as a function of cycling, the QPC model was applied to the experimental HRS curves after reset operation as shown in Figure 8.26. According to the QPC model, the current of a single filament is defined by $I = G_0 V$, as illustrated by the green dashed line.

The QPC parameters (ϕ, α and β) were extracted as illustrated in Figure 8.27(a) and (b). Alpha and phi show opposite evolutions, and their product remains constant as illustrated in Figure 8.27(c). The correlation between these two parameters can explain the current stability during cycling. Beta remains constant around 0.85 locating the constriction point near the bottom contact.

The reduction of the cell-to-cell variability by applying the ISPVA makes it a mandatory approach for programming operations. In cycling, the compensation effect between QPC parameters (ϕ and α) seems to be responsible for the current stability. The good performance of the devices, programmed using the ISPVA, makes thus ReRAM-based ELT approach a good candidate to be a Rad-Hard Non-Volatile Memory solution.

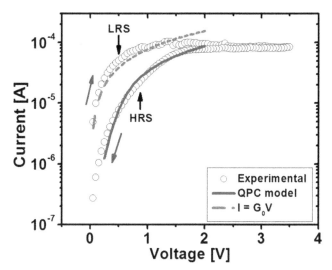

Figure 8.26 Reset I–V characteristics measured in DC mode. The blue line represents the adaption of the QPC model detailed by Equation (8.1).

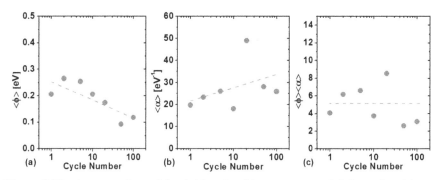

Figure 8.27 Average values of the ϕ (a) and α (b) parameters and their product (c) as a function of cycling.

8.4 Conclusions

In this chapter, several non-volatile memory device concepts that have been proposed in the last few years are reviewed. Among them, resistive RAM, ReRAM, technology is seen as a major candidate for Radiation-Hard memory devices. The reliability issues of 1T-1R-based HfO_2 ReRAM arrays are discussed in detail.

The switching performance of 1T-1R devices based on ELT as the selector device has been investigated in AC and DC modes, as well. Because of the high cell-to-cell variability in the DC mode, it is not possible to define a clear gap between HRS and LRS. The reduction of the cell-to-cell variability by applying the ISPVA makes it a mandatory approach for programming operations, thus increasing the feasibility in ReRAM usage.

References

[1] S. Gerardin, A. Paccagnella, Present and Future Non-Volatile Memories for Space, IEEE Trans Nuclear Science, vol. 57, no. 6, pp. 3016–3039, Dec. 2010.

[2] E. Ipek et al., Dynamically replicated memory: building reliable systems from nanoscale resistive memories, Proc ASPLOS, 2010. pp. 3–14, March 2010.

[3] R. Micheloni et al., "Non-Volatile Memories for Removable Media," Proceedings of the IEEE, vol. 97, no. 1, pp. 148–160, Jan. 2009.

[4] S. Banerjee, D. R. Chowdhury, Built-in self-test for flash memory embedded in SoC, IEEE DELTA 2006., pp. 379–384, Jan. 2006.

[5] R. Waser (Editor), Nanoelectronics & Information Technology, Wiley VCH 2005.

[6] A. Sheikholeslami, P. Gulak: A survey of circuit innovations in Ferroelectric random-access memories, Proc IEEE, Vol. 88, No. 3, pp. 667–689, 2000.

[7] J. M. Slaughter et al., J. Superconductivity: Incorporating Novel Magnetism, 15 (2002).

[8] Johan Åkerman, "Toward a Universal Memory", Science, Vol. 308. no. 5721 (22 April 2005), pp. 508–510.

[9] "Past, Present and Future of MRAM", NIST Magnetic Technology, 22 July 2003.

[10] C. H. Sie, A. V. Pohm, P. Uttecht, A. Kao and R. Agrawal, "Chalcogenide Glass Bistable Resistivity Memory", IEEE, MAG-6, 592, September 1970.

[11] D. Ielmini et al., IEEE Trans. Electron Dev. vol. 54, 308–315 (2007).

[12] R^{2}RAM Project (Radiation Hard Resistive Random-Access Memory), European Union's Horizon2020 Programme, www.r2ram.eu

[13] ObservatoryNANO, Briefing No. 4, ICT, Universal Memories.

[14] S.-S. Sheu et al., A 4Mb embedded SLC resistive-RAM macro with 7.2ns read-write random-access time and 160ns MLC-access capability, ISSCC, 2011.

[15] Y. S. Chen et al., Tech. Dig.-Int. Electron Devices Meet. (IEDM), 717, 2011.

[16] IHP Gmbh – Innovation for High Performance Microelectronics, Frankfurt (Oder), Brandenburg, Germany, https://www.ihp-microelectronics.com/en/start.html

[17] Ch. Walczyk et al, Pulse-induced low-power resistive switching in HfO_2 metal-insulator-metal diodes for nonvolatile memory applications, J. Appl. Phys., 105, 114103, (2009)

[18] Ch. Walczyk et al., "Impact of Temperature on the Resistive Switching Behavior of Embedded HfO_2-based RRAM devices", IEEE Trans. Electron Devices, 58, 3124, 2011.

[19] D. Walczyk et al., "Resistive switching characteristics of CMOS embedded HfO2-based 1T1R cells", Microelectronic Eng. 88, 1133, 2011.

[20] C. Zambelli et al., "Analysis and optimization of erasing waveform in phase change memory arrays", ESSDERC '09. pp. 213–216, 2009.

9

New Generation of NVMs Based on Graphene-related Nanomaterials

Paolo Bondavalli

Thales Research and Technology, France

"It is impossible to live in the past, difficult to live in the present and a waste to live in the future."

Frank Herbert, Dune

Non-volatile memories (NVM) are computer memory that can get back stored information even when not powered. Recently, NVMs have been identified as a key device for the future of computational system considering the important reduction of energy consumption compared to volatile memories (that do not store information after being turned off). To date, a "universal" new technology has not been identified considering the issues related to scalability and cost. Graphene-related materials offer an interesting and extremely promising alternative to existing technologies mainly based on Resistive Random Memories (ReRAM), which are memories exploiting the change of the resistance of a material under the effect of an electric field as an information write/erase principle for non-volatile data storage. ReRAM are fabricated using metal oxide materials as the sensitive materials or Flash memories exploiting the charge modulation of a transistor channel by a floating gate. In this chapter, we show the main scientific works concerning the properties memories based on graphene layers, graphene-oxide and reduced graphene oxide. These works have recently been issued but already show promising results especially in terms of writing time and endurance/cyclability compared to existing technologies. Moreover, these materials can potentially allow developing a highly scalable CMOS compatible technology.

This memory technology has two main advantages that can be exploited for space applications. They are not sensitive to radiation, or at less significantly less than flash memories, and can be easily reprogrammable to correct errors. Indeed, ReRAM memories are not sensitive to ionizing radiation and ElectroMagnetic Interference (EMI) as compared to charge accumulation of Flash memories. The different resistive switching (RS) mechanisms have reshuffled all the concepts of destructive and non-destructive single event effects in semiconductors, so critical in the space field. We can mention that carbonaceous materials are easily available compared to rare earths, used in existing memories and frequently found mixed with potentially dangerous radioactive ores such as thorium, and compatible with flexible substrate allowing potentially the implementation of new functionalities.

9.1 Introduction

Graphene is a one-atom-thick layer of carbon atoms arranged in a hexagonal lattice. It has been defined as the wonderful material, considering its properties in terms of mobility of charges, mechanical strength and thermal conductivity, which can pave the way for a new generation of devices. Graphene is the base for other carbon allotropes like nanotubes (CNT), fullerenes or graphite. In fact, many of the electronic and structural properties of the latter can be derived from graphene. Until about a few years ago, this two-dimensional form of carbon was known only in the three-dimensional stack of graphite or tightly bound to the surface. This is logic considering that this is natural form of this material. However, scientists from Manchester University, Andre Geim and Kostya Novoselov, famously cleaved it from the graphitic stack in 2004 (exploiting the so-called "scotch tape technique") and revealed the impressing electronic properties of this nanomaterial [1–3], granting them the 2010 Nobel Prize in Physics. This contributes to enhance the interest for this new nanomaterial, and on the 2D materials in general, amongst the scientific community. The graphene potentialities are attracting several researchers to probe opportunities in several directions in the "more Moore" or "beyond CMOS" optics such as novel graphene-based transistors or spintronics [4–6]. One of the last tendencies is to employ graphene and other related nanomaterials to fabricate non-volatile memories exploiting their "memresistive" behaviour storing a value of electrical resistance in a permanent way (i.d. in a non-volatile way). This occurs when a current is passed through these materials, their level of resistance to electricity (in Ohms) changes in a non-volatile and energetically "friendly" way (it lasts when the power is turned off, in opposite to volatile memories). Therefore, resistive memory exploits the change of the resistance of a material under the effect of an electric field as an information write/erase principle for non-volatile data storage. The reading of resistance states is non-destructive, and the memory devices can be operated without transistors in every cell [7, 8], as for flash type memories [9–14] (see Section 9.2.2), thus making a crossbar structure feasible. To be more precise, this kind of memories is called resistive random-access memory (RRAM or ReRAM) and is only one of the possible types of non-volatile way to store information in a permanent way. One of the most important advantages of graphene-related materials, compared to the previous materials, is that these materials can be implemented into flexible electronics [15–19], in the form of one thick atom layer as for graphene or in the form of layers of flakes of graphene oxide or reduced graphene oxide (see Section 9.2.2) and potentially can lower the final cost of the final device. Another great advantage of ReRAMs is its potential to implement them exploiting only two terminals to work (two contacts and not three as a common transistor, which has drain, source and gate, e.g. flash type memories), which could dramatically reduce the circuitry and allows the easier implementation in 3D architectures. Finally, the circuit commonly used for computer memory using graphitic layers-based inks [20] could be potentially easily printed or deposited (e.g. by spray-gun) on plastic sheets and used wherever flexibility is needed, such as in wearable electronics. This can be implemented using roll-to-roll fabrication technique. Potentially, this can influence a large field of application, with a very large market, as health monitoring [21–25], intelligent packaging [26, 27] (e.g. food packaging), card labels, badges, value paper, medical disposables and so on. Another potential huge market is the Radio-frequency identification (RFID). RFID is a technology to electronically record the presence of an object using radio signals. Indeed, an innovative alternative pathway to reduce RFID costs and integrate a memory chip to store data is to eliminate the silicon substrate completely and produce RFID and memory on the same flexible plastic substrate as the antenna [28–30]. Thanks to graphitic layers, the antenna and chip can be built on the same low-cost substrate and attachment costs can be removed.

9.2 Graphene-based Non-volatile Memories

In this section we present the main results concerning graphene and related nanomaterials (Graphene Oxide, i.d. GO, and Reduced Graphene

Oxide, i.d. R-GO) for the fabrication of non-volatile resistive memories (memresistive-like memories). While outlined in the International Technology Roadmap for Semiconductors (ITRS) 2011 [31] chapter concerning Emerging Research Devices (i.d. ERD), the interest is quite recent. It is specifically highlighted that "ultrathin graphite layers are interesting materials for macromolecular memories thanks to the potential fabrication costs that are considered as the primary driver for this type of memory, while extreme scaling is de-emphasized". However, ITRS outlines that the memory operation mechanisms and the physics, as for more generally Resistive Random-Access Memories (i.d. ReRAM), are still unclear and that a deeper research in this field is necessary to improve the comprehension of the phenomenon and the efficiency of the devices. For the sake of the coherence of the chapter, this contribution will not concern the fabrication of carbon nanotubes (i.d. CNTs)-based non-volatile memories, as shown in some past interesting papers. Indeed, in case of CNTs, which are rolled up foils of graphene, the physical phenomena exploited were different, such as the contact between two CNTs as a mechanical switch, as pointed out by Rueckes et al. [32], the local change of sp^2 structure (orbital hybridisation) by local pulse [33], the charge trapping in CNTs [34] or the photo-activation [35–37]. These are not the same physical mechanisms exploited in memories based on graphene-related materials. These last will be developed in the next paragraphs.

9.2.1 Graphene and Graphitic Layers

The first paper pointing out the potential for the utilization of graphene for resistive non-volatile memories exploiting two terminal structures was issued in 2008 by Stadley et al. at Caltech [38]. Stadley and co-workers reported the development of a non-volatile memory element based on graphene break junctions. These junctions were simply obtained achieving two-terminal devices, transferring graphene sheets on SiO_2/Si substrate and depositing metal electrodes on them using e-beam lithography process. Indeed, after applying a specific voltage under ultra-vacuum condition (10^{-7} Torr), they were able to create a breakdown in the graphene layer (see Figure 9.1(a)). In these papers, all the measurements were performed under ultra-vacuum to avoid environmental gas interferences.

However, after repeating the voltage cycles more number of times, they demonstrated that the resistance value changed as a function of the voltage (see Figure 9.1(b)): scientists were able to move from a high-resistance state (Off state) to a lower one as a function of the voltage (On state). This effect

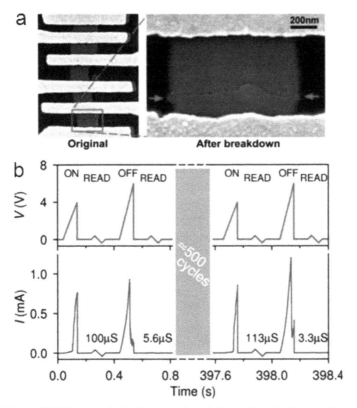

Figure 9.1 (a) SEM image of the device before (left panel) and after breakdown (right panel). The arrows indicate the edges of the break of the graphene layer. (b) Repeatable programming over hundreds of cycles. Top left panel: Voltage applied to the junction vs time. A ramp with a peak value of $\sim$4 V corresponds to an ON pulse, while a ramp with a peak value of $\sim$6 V corresponds to an OFF pulse. A small sawtooth-shaped read-out pulse is applied after each write to determine the junction conductance. Bottom left panel: current flow through the junction, with the low-bias conductance labelled above each read-out pulse. Right panels: Similar data taken after approximately 500 intervening cycles, demonstrating the reproducibility of the switching behaviour [38].

was non-volatile, and the devices demonstrated thousands of writing cycles (10^5) and long retention times (more than 24 h to, perhaps, indefinitely at room temperature).

The observed switching behaviour of graphene was impressing in terms of reproducibility and non-volatility. The more plausible explanation was that the conductance occurred along a small graphene ribbon that bridges the contacts. The absence of a gate voltage effect excluded switching mechanisms arising from the storage of charge in traps in the oxide layer or from the formation of a small metallic island that may function as a quantum dot. To deeply understand the underlying physical mechanism, researchers at Caltech investigated the time-resolved behaviour of the switching from the OFF to ON states, shown in Figure 9.2(a). The conductance I/V showed well-defined steps, with magnitude $\sim G_Q$. Here, $G_Q = 2e^2/h$, the conductance quantum, where "e" is the electron charge and "h" is Planck's constant. Since G_Q is the conductance of a spin-degenerate one-dimensional conductor, e.g., a linear chain of gold atoms [39], observation of steps in the conductance suggests that the device conductance states are likely multiples of highly transmitting quantum channels. Therefore, authors proposed a model for device operation based on the formation and breaking of carbon atomic chains that bridge the junctions (see Figure 9.2(a) – right). This mode and the results seem to put in evidence the potential for multi-resistive states.

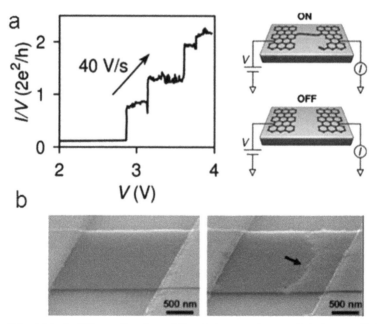

Figure 9.2 (a) Left: I/V as a function of V pointing out the quantic behaviour linked to the formation of atomic chains. Right: Proposed physical mechanism at the base of the non-volatile resistive behaviour (formation of atomic carbon chains) [38]. (b) Tilted-view SEM images of two different devices before and after applying the voltage difference necessary to break the layers defined as V_{break}. The arrow shows the fracture across the stripe due to V_{break} [39].

The same physical explanation can be evoked for the results obtained in 2009 by Sinitski et al. [40] at Rice. In this case, they tested similar carbon-based structures (see Figure 9.2(b)). The sensitive layer was not composed by more staked graphitic discs, reaching 10 nm of thickness, grown by CVD (Chemical Vapour Deposition) process. These discs were constituted by discontinuous graphitic sheets of around 5 nm in length and extensively buckled. Researchers evaluated the voltage necessary to break the layer. After calculating the Joule heating, they obtained the breakdown voltage, which is the applied voltage difference necessary to break the layer, as:

$$V_{break} \, \alpha \, [(C\rho\gamma T_{break})/\tau]^{1/2} \, 1 = A \qquad (9.1)$$

where C, ρ and γ are, respectively, the specific heat, the density of CVD grown graphitic sheets and resistivity. T_{break} is the temperature of breakdown of the graphitic layer. "A" is independent of the devices and "l" the length of the device. The main conclusion is that the voltage necessary to break the layer is directly proportional to the length. This is very interesting considering that reducing the dimensions will allow to reduce the energy consumption to write the information.

Using a different approach, in 2009, the same team from Rice University [41] had grown graphitic layers on a freestanding silicon oxide nanowire.

Indeed, they used the nanowire to preserve the mechanical integrity of the device after applying a specific voltage that broke the graphitic layers (as in case of planar junctions). In case of the utilisation of CNTs, breaking the graphitic layers also meant to break the nano-object definitively, without any possible reversibility of the phenomenon. The team from Rice fabricated two terminal devices consisting of discontinuous 5–10 nm thin films of graphitic sheets grown by chemical vapour deposition on nanowires possessing stable, rewritable, non-volatile and non-destructive read memories with on/off ratios of up to 10^7 and switching times of up to 1 μs (tested limit).

All the samples were not tested at ambient temperature but in ultra-vacuum condition (5 $*10^{-5}$ Torr). One of the most important results was the extremely good stability of the performances after several days and after X-ray exposure. The only concern about this kind of devices is that they present all the problems of fabrication and manipulation of typical of nano-size objects. For this reason, from an industrial point of view, their exploitation is quite difficult.

Another interesting work by Wu et al. [42] in 2012 dealt with the electrical properties of graphene sheet suspended on a patterned indium tin oxide (ITO) electrode pair. They measured very interesting on/off ratio of 6 orders and retention time of at less 10^4 s in ambient conditions. In this case, data storage can be achieved by applying voltage bias and rewritten after a simple heat treatment. In this case, scientists did not observe crack in the graphitic flake and evoked the local oxidation of graphene sheet at the graphene/electrode interface as the main phenomenon for switching. It was highlighted that when the measure was performed in nitrogen atmosphere, there was no switching phenomenon. The authors stated that the switch happens only in ambient atmosphere when oxygen changed the injection barrier at the contact. The degassing using a thermal process allowed coming back to the high conduction state.

9.2.2 Non-volatile Resistive Memories Based on GO and R-GO Oxide Layers

Another approach followed by some teams has been to exploit the "memristive" effect in graphene oxide (e.g. GO) layers or flakes in order to achieve non-volatile memories. Indeed, GO is a wide band gap material (6 eV) with potential for modulation thanks to oxidation/reduction process providing tunability of the electronic, mechanical and optical properties. GO [43] is commonly obtained by oxidation of graphite using modified Hummer's method, where the long oxidation time is combined with a highly effective methods for the purification of reaction products [44] (Figure 9.3). A great advantage of this material is the cost, which is 1/10 of the price of graphene flakes.

Moreover, thanks to its hydrophilic character, GO flakes can be put in stable solutions easily and deposited on large surface using simple methods as spray coating or drop casting. In this case of the utilization of GO, we will

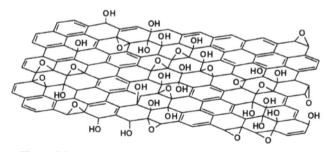

Figure 9.3   One of the graphene oxide typical structures [43].

exploit another configuration, which is similar to crossbar memories [45–47] (vertical junctions composed of piling up the different layers, sandwiching the active one as Metal–Insulator–Metal (MIM)-type structures). In this context, we mention the work of He et al. in 2009 [46] where reliable and reproducible resistive switching behaviours were observed in GO thin films prepared by vacuum filtration method, the common technique used to fabricate CNT-based buckypapers [48]. They fabricated Cu/GO/Pt structures showing an on/off ratio of about 20, a retention time longer than 10^4 s and a switching threshold voltage of less than 1 V. The suggested physical principle at the origin of the switching effect was the desorption/absorption of oxygen-related groups on the GO sheets as well as the diffusion of the top electrodes. These experiments underlined that GO was potentially useful for future non-volatile memory applications. One of the more important works on this material was issued in 2010 by Jeong et al. [49, 50]. This team demonstrated the non-volatile effect on the resistance of a 70 nm layer thick GO in a layered structure composed of Al/GO/Al (Figure 9.4) in a crossbar configuration.

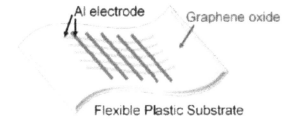

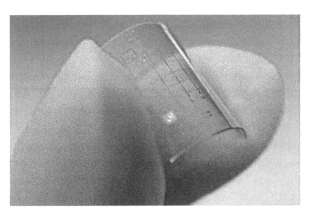

Figure 9.4 Top: Schematic illustration of a G-O-based flexible crossbar memory device. Bottom: Real crossbar memory devices based on GO [49].

Jeong et al. were one of the first teams that pointed out the performances of these kinds of layers on flexible substrates. The researchers form KAIST demonstrated that using GO layers they were able to achieve flexible non-volatile resistive memories with impressive performances under traction and compression (they kept three orders between on (LRS) and off (HRS) levels). Concerning the physics explanation of the phenomenon, Hong et al. [51, 52] performed a deep analysis of the switching mechanism for this kind of devices. Hong and co-workers underlined that these structures showed per-formances depending on the origin of the top contacts. For example, in case of Au-based top electrodes, there was no oxygen migration in opposition to Al electrodes. The effect of bottom contacts was also quite important. Indeed, if the roughness was too high, the GO layer had such cracks and rough surface that the top electrode material could easily penetrate and then build filaments that hindered the switching operations. This drastically reduced the device lifetime (only 100 cycles) because of the formation of a permanent conductive path between the two electrodes. These results were highlighted by X-ray photoelectron spectroscopy (XPS) measurements that pointed out the permanent presence of Al near the bottom electrode as the main raison for failure. Evaluating the effect of the migration of oxygen versus the formation of Al-based filaments, Hong and co-workers stated that the formation of conducting filaments was a local phenomenon and that the oxygen migration was the dominant mechanism. Indeed, they discovered that the leakage current between the bottom and lower electrodes was dependent on the cell dimension. In fact, when the cells were larger, the effect of the oxygen migration or reduction on the conduction was enhanced, proportion-ally to the surface. To avoid the failure mechanism linked to the formation of permanent conductive filaments through the material cracks, one suggestion is to exploit deposition techniques that allow a more uniform distribution of the deposited material. One technique is the deposition by spray coating heating the substrate to avoid the so-called "coffee-ring effect" [53, 54], which is responsible for the non-uniformity of the deposition [55, 56]. This deposition could potentially prevent the formation of cracks linked to the roughness in the GO layers. Regarding GO layers, Panin et al. [57] in 2011 used a different approach. Indeed, they fabricated planar devices depositing flakes of GO using spray coating technique on patterned substrates. The GO flakes had the dimensions of several tens of nanometres. The distance of the pre-patterned electrodes was around 20 μm. Panin and co-workers observed after applying a forming voltage of 5 V, a sharp unipolar resistive switching. The on/off ratio was around 10^3 as in the other works (Figure 9.5).

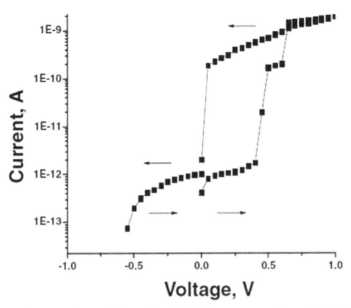

Figure 9.5 Resistive switching of the Al/GO/Al structure pre-formed at 5 V [57].

These measurements were performed at room temperature. The local changes of resistance in the layer near the Al/GO interface under bias voltage as a result of sp^3 to sp^2 structure reconstruction are assumed to be responsible for the observed resistive switching. The mechanism of the resistive switching in the structure is supposed to be related to the electrodiffusion of oxygen and the formation of sp^2 graphene clusters in the sp^3 insulating graphene oxide layer formed near the electrode by a pre-forming process.

Another important study was performed by Vasu et al. at Bangalore University (India) in 2011 [58]. They reported very simple unipolar resistive switching device using ultrathin (~20 nm) films of reduced graphene oxide (e.g. R-GO) with on/off ratios reaching 5 orders of magnitude. The thin films were formed at room temperature simply by drop-casting R-GO from suspension on ITO glass electrode, followed by aluminium or gold deposition (no difference was pointed out after the measurements). A very interesting result was also the switching time that could attain 10 µs with an on/off ratio of 100. In this case, the formation of nano-filaments of carbon atoms was evoked as main physical working principle.

In Table 9.1, the main results present in the literature obtained using the targeted materials are summarized.

Table 9.1 Main results of resistive non-volatile memories built with graphene, GO, R-GO

Source	Material and Configuration	On/off Max	Switching Time	Cyclability	Retention Time	Character	Energy Consumption to Write (J)
[Li08]	Graphitic layers on nanowire (planar)	10^7	1 μs	At less 1000 with no variation	1 day at less (15 days of consecutive switching)	Unipolar	~4 × 10^{-15}
[Stadley08]	Graphene (planar)	10^2	100 ms	10^5 without visible changes in signals	To be performed in more than 24 h	Unipolar	Not clear
[Sinitski09]	Graphene (planar)	10^7	1 μs (tested limit)	at less 22,000 with no variation	Not specified, even if the device is defined very stable	Unipolar	~8 × 10^{-17}
	Carbon (vertical)	10^6	1 μs (tested limit)	at less 200	Not specified, even if the device is defined very stable	Unipolar	~7 × 10^{-17}
[He09]	GO (vertical)	20	No data	at less 100	10^4 s	Bipolar	No switching time
[Son10]	Organic bistable devices	10^7	No data	10^7	No data	Bipolar	No switching time

[Jeong10]	GO (vertical)	10^3	No data	at less 100	10^5 s	Bipolar	No switching time
[Hong11]	GO (vertical)	10^3	No data	100 and failure	Not specified, even if the device is defined very stable	Bipolar	No switching time
[Panin11]	GO (planar)	10^3	No data	No data	No data	Bipolar and Unipolar	No switching time
[Vasu11]	RGO (vertical)	10^5	$10\ \mu s$	Few hundreds	Not specified, even if the device is defined very stable	Unipolar	$\sim 10^{-13}$
[Wu12]	Graphene sheet between two metal electrodes	10^6	No data	No data	10^4	50 ms but limited by the experimental setup	Switching time not correctly measured ($< 50 \times 10^{-9}$)

9.3 Other Approaches to Achieve Non-volatile Memories Using Graphitic Layers

In the previous paragraphs, we have discussed about resistive non-volatile memories. Other teams have decided to use different approaches to exploit the properties of graphene-based layers to fabricate non-volatile memories. We will not discuss about mechanical switches fabricated moving suspended graphitic layers as mentioned by Milaninia et al. [59] and by Kim et al. [60] because they cannot be used as non-volatile memories. As it will be shown in the next paragraph, some teams have chosen to use the hysteretic electrical behaviour of graphene, or multilayered graphene, when used as channel in a Field Effect Transistor (FET)-type configuration. This is mainly based on the chemical modification of graphene under an electric field that will enhance environmental interactions. However, the more promising results, in terms of performances as well as potential industrial exploitation, have been obtained in case of the fabrication of Flash-type memory [61–65] using graphene or multi-layered graphene (MLG) as the floating gate (FG). Basically, a flash memory is an electronic non-volatile computer storage medium that can be electrically erased and reprogrammed. In flash memory, each memory cell resembles a standard MOSFET [66], where the transistor has two gates instead of one. Indeed, there is the traditional top gate, also called control gate, and the so-called floating gate embedded in an oxide layer, and so electrically insulated. In the FG, which is located between the top gate and the MOSFET channel, any electrons placed on it, activating the top gate, are trapped and will not discharge for many years. If the FG holds a charge, then it screens (partially cancels) the electric field from the top gate and therefore modifies the threshold voltage (V_T) of the MOSFET. This operation is non-volatile and is reversible if another adequate voltage is applied on the top gate. The present technology used for flash-type memories, employs doped polysilicon as material for the FG. One of the main advantages of graphene/multi-layered graphene (MLG) is the reduction of voltage to achieve a correct memory window, which is the shift of the threshold voltage of the transistor when the memory is switched from 0 to 1 state. In the present flash memories, a voltage difference of ± 20 V [67] is necessary to program/erase to achieve a memory window of 1.5 V, which is the industrial standard. Some works have demonstrated, as shown in the following paragraphs, that thanks to the higher density of states (DOS) of graphene compared to degenerately doped polysilicon, this difference can reduced up to 6 V. Another advantage is the higher work function of graphene [68–71] that is directly linked to the larger barrier height between the floating gate and the oxide, which embeds the FG. This potentially allows achieving longer retention time, potentially more than 10 years. Finally, there is a possibility to produce devices using 2D devices with a lower dimensionality that will potentially allow an easier 3D implementation and so a higher memory density.

9.3.1 Graphitic-based Non-volatile Memory Using a Transistor Configuration

One of the first papers highlighting the interest of using graphene-based Field Effect Transistors (FET) for non-volatile memories was issued by Wang et al. at the Nanyang Technological University in Singapore in 2010 [72]. Indeed, they observed exactly the same phenomenon observed in case of FET with one carbon nanotube as channel: a hysteretic phenomenon where they observed the formation of a loop of the current value between drain and source, I_{ds}, as a function of the gate voltage, V_{gs}, during the voltage sweep cycle from negative to positive values of V_{gs} and back. This phenomenon is caused by the presence of moisture interacting with the transistor dielectrics. This effect, as it has been demonstrated, can be removed in vacuum [73]. Moreover, the switching time in case of activation of the hysteretic phenomenon in FET using graphene is of the order of some seconds, which is too high for memories applications.

In 2008, pioneering works of Echtermeyer et al. at the Advanced Research Center in Aachen (Germany) and of A.K. Geim at the University of Manchester (UK) were performed adopting another point of view compared to the previous studies [74,75]. Considering the difficulty to achieve nanoribbons with dimensions smaller than 10 nm to open a gap, they exploited another phenomenon. Indeed, they fabricated two-gate transistor structures as shown in Figure 9.6, where the graphene has a width of 10 μm and was covered by a SiO_2 layer.

They observed that applying a voltage spanning from –5 to +5 V, for positive voltage, there were a strong enhancement of current of 5/6 order, and so a reduction of the resistance that was non-volatile. Echtermeyer and co-workers suggested that, considering that the measurements had been performed in ambient conditions, the water molecules were split in H+ and OH^- (see Figure 9.7) by the electric field and tended to attach to the graphene, creating graphene or graphene oxide (that are insulating materials), passing through the highly porous silicon dioxide layer surface, and so drastically changing the conductivity of the channel in a non-volatile way. This phenomenon allowed obtaining a high on/off ratio, quite unusual for graphene (and not

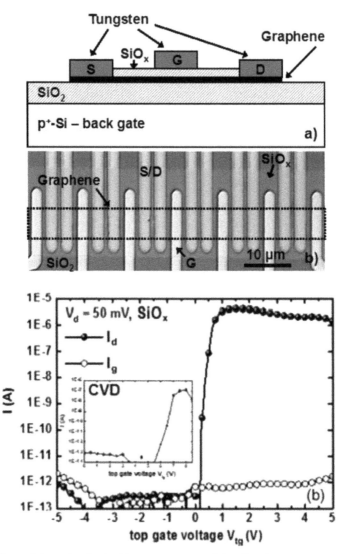

Figure 9.6 (a) Schematic of a double-gated graphene field effect device (FED). (b) Current between drain and source as a function of the top gate voltage value [74, 75].

graphene nanoribbons)-based FET. However, even if the results seem to be very promising, more studies must be done on the switching time, which intuitively appears to be quite long considering the effects involved. We can add that the graphene layer have been obtained by mechanical cleavage,

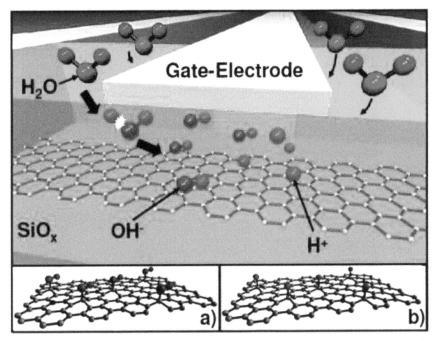

Figure 9.7 Schematic view of the electrochemical modification of graphene. As a function of polarity, the reactive ions diffuse towards graphene and chemically change the electrical character of graphene, creating derivative materials such as graphene (inset b) or graphene oxide (inset a), [74, 75].

which is a technique that allows fabricating high-quality samples but that it is not industrially suitable. It is necessary to test this approach, for example, using CVD graphene in order to achieve a potential parallel fabrication of this kind of devices. It has also to be verified if it is fundamental to obtain a single graphene layer or if more stacked layers of graphene can reach the same performances.

9.3.2 Non-volatile Flash-type Memories Based on Graphene/Multi-layered Graphene

Another potential utilization of graphene is as the floating gate in flash-type memories. An extremely innovative approach has been proposed by Bertolazzi et al. of the Ecole Polytechnique de Lausanne (EPFL) in Switzerland. They were able to fabricate a flash memory using the same architecture as the existing ones, exploiting only 2D materials [76].

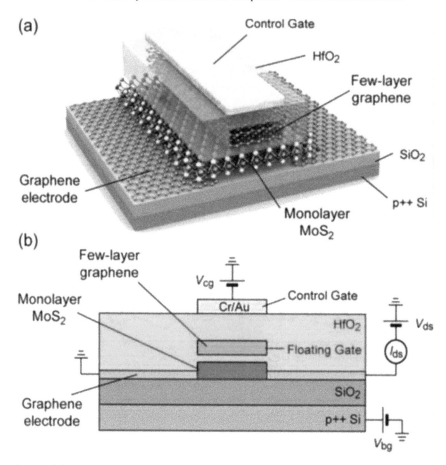

Figure 9.8 MoS$_2$/graphene heterostructure memory layout. (a) Three-dimensional schematic view of the memory device based on single-layer MoS$_2$. (b) Schematic view of the flash-type memory cell with a single-layer MoS$_2$ acting as semiconducting channel and graphene contacts and multilayer graphene (MLG) as the floating gate [76].

Bertolazzi and co-workers were able to build a flash-type memory using MoS$_2$ as the channel and few graphene layers as the floating gate (FG) (see Figure 9.8), embedded in HfO$_2$ oxide, where the charge is trapped. It is clear that this kind of structure is extremely original, but some performances especially considering the switching time (the lowest value is 100 ms) are not adequate. Some doubts can be raised on the fabrication technique, which includes three quite complex transfer steps that are not suitable now for an industrial exploitation. Hong et al. at IBM in Watson (US) fabricated a

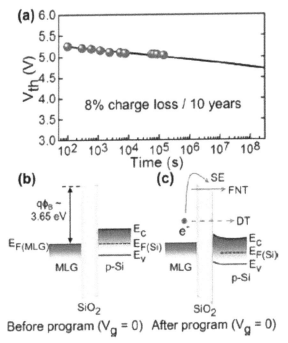

Figure 9.9 Retention characteristics of Graphene Flash Memory (GFM). (a) Retention measurement of MLG-FM showing only 8% of charge loss in 10 years at room temperature. (b) Energy band diagram of MLG/SiO$_2$/Si junctions before programming. (a) Energy band diagram after programming, which shows three possible mechanisms for charge loss during retention state: SE, FNT and DT. All three mechanisms depend exponentially on the barrier height (φ_B) between the work function of graphene and the SiO$_2$ tunnel oxide [77].

flash memory [77] using some graphitic layers, grown by CVD on Si, as the FG. In this case, the technological approach appears potentially scalable, CMOS-compatible and so industrially suitable (see Figures 9.9 and 9.10) compared to the previous one. IBM researchers observed a potential retention time of 10 years, with a loss of only 8% of the charges in the FG, but also simulating the potential crosstalk between neighbour cells, stated that this kind of memories showed negligible interference down to 10 nm (in case of common polysilicon FG, the interference raises dramatically under 25 nm). Briefly, considering the power reduction (as told previously, thanks to the higher DOS compared to common FG fabricated using polysilicon) and the increase in the storage density, Hong and co-workers estimated a potential reduction of 75% in the operating energy of this kind of memory. However,

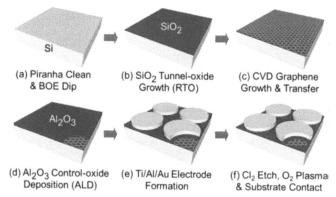

(a) Piranha Clean & BOE Dip

(b) SiO₂ Tunnel-oxide Growth (RTO)

(c) CVD Graphene Growth & Transfer

(d) Al₂O₃ Control-oxide Deposition (ALD)

(e) Ti/Al/Au Electrode Formation

(f) Cl₂ Etch, O₂ Plasma & Substrate Contact

Figure 9.10 Technological steps for the fabrication of flash-type memories using graphene as FG [77].

in this work, there are no data on the switching time and therefore we cannot compare their performances with existing memories.

9.4 Conclusions

Graphene-based memory is a completely new technology that needs more specific developments and time to demonstrate its utility especially in the challenging field of NVM for space application. This memory could constitute a real breakthrough compared to existing technologies also considering the dramatic potential reduction of energy consumption. These memories can be developed to bear harsh environments with a focus on radiation-resistant components. The stability (also in temperature), resilience, non-volatility and on/off ratio that in certain cases can attain 10^7 or the switching time of 1 μs (competitive with flash-type memories), potential switching voltages of 3–4 V, combined with predictable fabrication-controlled I–V behaviour, simple two-terminal geometry (no need for a gate and so dramatically reducing circuitry) and access to mass fabrication for the two approaches make them extremely attractive structures for non-volatile memories.

Indeed, thanks to thin-film technologies, electronic functionality can be foreseen in very large quantity and at very low cost on substrates such as plastic and paper. Additional functionality is also an attractive feature of carbon-based memories. For example, by using these kinds of materials, we open the route for memories on flexible substrates, a key building block to enable the success of flexible electronics. Carbon-based resistive memories should also offer the capability for multi-level storage and "memristive-like" behaviour, as seen in other resistive memory materials. Multi-level storage allows for the storage of more than one bit per cell, so increasing data storage densities, while memristive-like behaviour can be exploited to provide a remarkable range of signal processing/computing-type operations, including implementing logic, providing synaptic and neuron-like "mimics" and performing, in a very efficient way, analogue signal processing functions (such as multiply-accumulate operations).

In case of graphitic-based memories based on other approaches, the transistor-type configuration exploiting the hysteretic effect, in our opinion, has a limited potential compared to new flash-type memories using graphene or MLG as the FG, because their architecture can be unlikely miniaturized and their performances especially considering the switching time are difficult to evaluate. In case of flash-type memories, these last exploit intrinsically characteristics of the materials such as the higher DOS (which allow reducing the voltage for the memory window) or the higher work function (which allows reaching more than 10 years of charge storage) that really allows improving the performances of this kind of memories. Moreover, it has been demonstrated to achieve high-density architecture with limited crosstalk. It is necessary to make an effort on the miniaturization of these devices, but the recent works, e.g. [20] make us think that this technology can be quite rapidly in competition with existing flash memories, also because it is CMOS-compatible and easy to implement in existing production lines.

References

[1] A. K. Geim and K. S. Novoselov, Nature Materials, 6, 183 (2007).

[2] K. S. Novoselov, Q3 358 Physics World, 22(8), 27 (2009).

[3] K. S. Novoselov, A. K. Geim, S. V. Morozov, 359.

[4] Tombros, N., Jozsa, C., Popinciuc, M., Jonkman, H. T. & van Wees, B. J. Electronic spin transport and spin precession in single graphene layers at room temperature. Nature 448, 571–574 (2007).

[5] Ohishi, M. et al. Spin injection into a graphene thin film at room temperature. Jpn. J. Appl. Phys. 46, L605-L607 (2007).

[6] Cho, S., Chen, Y. F. & Fuhrer, M. S. Gate-tunable graphene spin valve. Appl. Phys. Lett. 91, 123105 (2007).

[7] Lee M. J., Suh D. S., Seo S., Jung R., Ahn S. E., Lee C. B., Kang B. S., Ahn S. E., Lee C. B., Seo D. H., Cha Y. K., Yoo I. K., Kim J. S., Park B. H. 2007. Two Series Oxide Resistors Applicable to High Speed

and High Density Nonvolatile Memory, Advanced Materials 19 22 3919–3923 0935–9648.

[8] Waser R., Aono M., 2007 Nanoionics-based resistive switching memories, Nature Materials 6 11 833 840 1476–1122.

[9] R. Bez, E. Camerlenghi, A.Modelli and A. Visconti, Proceeding of the IEEE, 91, 4, 489, 2003.

[10] S. Aritome, "Advanced flash memory technology and trends for file storage application," in Tech. Dig. IEDM, 2000, p. 763.

[11] G. Muller, T. Happ, M. Kund, G. Yong Lee, N. Nagel, and R. Sezi, Status and Outlook of Emerging Nonvolatile Memory Technologies, In proceeding of: Electron Devices Meeting, 2004. IEDM Technical Digest. IEEE International.

[12] S. Lai, "Flash memories: Where we were and where we are going," in IEDM Tech. Dig., 1998, pp. 971–973.

[13] P. Pavan, R. Bez, P. Olivo, and E. Zanoni, "Flash memory cells-An overview," Proc. IEEE, vol. 85, pp. 1248–1271, Aug. 1997.

[14] P. Pavan and R. Bez, "The industry standard Flash memory cell," in Flash Memories, P. Cappelletti et al., Ed. Norwell, MA: Kluwer, 1999.

[15] Graphene-based flexible and stretchable thin film transistors Chao Yan, a Jeong Ho Cho*b and Jong-Hyun Ahn*a Nanoscale, 2012, 4, 4870.

[16] S. Bae, H. Kim, Y. Lee, X. Xu, J.-S. Park, Y. Zheng, J. Balakrishnan, T. Lei, H. Ri Kim, Y. I. Song, Y.-J. Kim, K. S. Kim, B. Ozyilmaz, J.-H. Ahn, B. H. Hong and S. Iijima, Nat. Nanotechnol., 2010, 5, 574–578.

[17] Y. Wang, Y. Zheng, X. Xu, E. Dubuisson, Q. Bao, J. Lu and K. P. Loh, ACS Nano, 2011, 5, 9927–9933.

[18] E. H. Lock, M. Baraket, M. Laskoski, S. P. Mulvaney, W. K. Lee, P. E. Sheehan, D. R. Hines, J. T. Robinson, J. Tosado, M. S. Fuhrer, S. C. Hern_andez and S. G. Walton, Nano Lett., 2011, 12, 102–107.

[19] Wei Xin, Zhi-Bo Liu, Qi-Wen Sheng, Ming Feng, Li-Gang Huang, Peng Wang,Wen-Shuai Jiang, Fei Xing,Yan-Ge Liu and Jian-Guo Tian, Flexible graphene saturable absorber on two-layer structure for tunable mode-locked soliton fiber laser, 2014 | Vol. 22, No. 9 | DOI:10.1364/OE.22.010239 | OPTICS EXPRESS.

[20] Inkjet Printing of High Conductivity, Flexible Graphene Patterns, E B. Secor, P L. Prabhumirashi, K Puntambekar, M L. Geier, and M C. Hersam J. Phys. Chem. Lett., 2013, 4(8), pp 1347–1351.

[21] Nature Communications Flexible polymer transistors with high pressure sensitivity for application in electronic skin and health monitoring, Gregor Schwartz, Benjamin C.-K. Tee, Jianguo Mei, Anthony L. Appleton, Do Hwan Kim, Huiliang Wang Zhenan Bao.

[22] Sekitani T., Zschieschang U., Klauk H. & Someya T. Flexible organic transistors and circuits with extreme bending stability. Nature Mater. 9, 1015–1022 (2010).

[23] Ramuz M., Tee B. C.-K., Tok J. B.-H. & Bao Z. Transparent, optical, pressure-sensitive artificial skin for large-area sretchable electronics. Adv. Mater. 24, 3223–3227 (2012).

[24] Wagner S. et al. Electronic skin: architecture and components. Phys. E 25, 326–334 (2004).

[25] Sokolov A. N., Tee B. C. K., Bettinger C. J., Tok J. B. H. & Bao Z. Chemical and engineering approaches to enable organic field-effect transistors for electronic skin applications. Acc. Chem. Res. 45, 361–371 (2012).

[26] L. Yam, Paul T. Takhistov, Joseph Miltz, Intelligent Packaging: Concepts and Application Kit Journal of Food Science, 70, 1, pages R1-R10, January 2005.

[27] Terho K. Kololuoma; Markus Tuomikoski; Tapio Makela; Jali Heilmann; Tomi Haring; Jani Kallioinen; Juha Hagberg; Ilkka Kettunen; Harri K. Kopola, Towards roll-to-roll fabrication of electronics, optics, and optoelectronics for smart and intelligent packaging, Proc. SPIE 5363, Emerging Optoelectronic Applications, 77 (June 25, 2004).

[28] Inkjet-printed antennas, sensors and circuits on paper substrate, S Kim, B Cook, T Le, J Cooper; H Lee, V Lakafosis; R Vyas; R Moro, M; Bozzi, A. Georgiadis, A Collado, M M. Tentzeris, IET Microwaves, Antennas & Propagation, Volume 7, Issue 10, 16 July 2013, p. 858–868.

[29] Le, T.; Ziyin Lin; Wong, C.P.; Tentzeris, M.M., "Novel enhancement techniques for ultra-high-performance conformal wireless sensors and "smart skins" utilizing inkjet-printed graphene," Electronic Components and Technology Conference (ECTC), 2013 IEEE 63rd, vol., no., pp.1640,1643, 28–31 May 2013.

[30] H.; Pham, D.T.; Xiaochuan Xu; Chen, M.Y.; Hosseini, A.; Xuejun Lu; Chen, R.T., "Inkjet-Printed Two-Dimensional Phased-Array Antenna on a Flexible Substrate," Antennas and Wireless Propagation Letters, IEEE, vol.12, no., pp.170,173, 2013.

[31] http://www.itrs.net/Links/2012Winter/1205%20Presentation/ERD_1205 2012.pdf

[32] Rueckes T, Kim K, Joselevich E, Tseng GY, Cheung CL, Lieber CM, 2000. Carbon Nanotube-Based Nonvolatile Random Access Memory for Molecular Computing, Science 7 289 5476 94–97.

[33] Kreupl F, Graham AP, Liebau M, Duesberg GS, Seidel R, Unger E, 2004. Carbon nanotubes for interconnect applications, IEEE International Electron Device Meeting Technical Digest, 683–686.

[34] Tour, J. M.; Zhong, L. Molecular Computing: Integration of Molecules for Nanocomputing, In Nanoscale and Bio-Inspired Computing, Mehrnoosh Eshaghian-Wilner, M., Ed. Wiley, 2009, 295–326.

[35] Agnus G, Zhao W, Derycke V, Filoramo A, Lhuillier Y, Lenfant S, Vuillaume D, Gamrat C, 2010. Two-terminal carbon nanotube programmable devices for adaptive architectures. Advanced Materials 22 702.

[36] Brunel D., Lévesque P. L., Ardiaca F., Martel R., Derycke V., 2008, Control over the interface properties of nanotube-based optoelectronic memory devices, Appl. Phys. Lett. 102, 013103.

[37] Zhao W, Agnus G, Derycke V, Filoramo2 A, Bourgoin2 and Gamrat C, 2010, Nanotube devices based crossbar architecture: toward neuromorphic computing, Nanotechnology 21 175202.

[38] Standley B., Bao W., Zhang H. Bruck J., Ning Lau C., Bockrath M. 2008. Graphene-Based Atomic-Scale Switches, Nanolett. 8, 10, 3345–3349.

[39] Agrait N, Levy-Yeyati A, van Ruitenbeek JM, 2003. Quantum properties of atomic-sized conductor, Phys. Reps. 377 81–380.

[40] A.Sinitskii, J.M.Tour, Lithographic graphitic memories, 3, 9, 2760 ACS Nano, 2009.

[41] Li Y., Sinitskii A., Tour J.M. 2008. Electronic two-terminal bistable graphitic memories, Nat Mat 7 966.

[42] Wu C, Li F, Zhang Y, Guo T 2012. Recoverable electrical transition in a single graphene sheet for application in nonvolatile memories, Appl. Phys. Lett. 100, 042105.

[43] The chemistry of graphene oxide Daniel R. Dreyer, S Park, C W. Bielawski and R S. Ruoff Chem. Soc. Rev., 2010, 39, 228–240.

[44] Hummers, W. S. & Offeman, R. E. Preparation of graphitic oxide. J. Am. Chem. Soc. 80, 1339 (1958).

[45] CMOS-like logic in defective, nanoscale crossbars, Greg Snider, Philip Kuekes and R Stanley Williams, Nanotechnology 15 (2004) 881–891.

[46] Complementary resistive switches for passive nanocrossbar memories Eike Linn, Roland Rosezin, Carsten Kügeler, Rainer Waser, Nature Materials 9, 403–406 (2010).

[47] He C. L., Zhuge F., Zhou X. F., Li M., Zhou G. C. 2009. Nonvolatile resistive switching in graphene oxide thin films, Appl. Phys. Lett. 95, 232101.

[48] Z. Wu, Z. Chen, X. Du, J.M. Logan, J. Sippel, M. Nikolou, K. Kamaras, J.R. Reynolds, D.B. Tanner, A.F. Hebard, Science 305 (2004) 1273–1276.

[49] Jeong H Y, Yun Kim O, Won Kim J, Hwang J O, Kim J-E, Yong Lee J, Yoon T H, Cho B J, Kim S O, Ruoff R S., Choi S-Y 2010. Graphene oxide thin films for flexible nonvolatile memory. Nanolett. 10 4381–4386.

[50] http://spectrum.ieee.org/semiconductors/nanotechnology/flexible-graphene-memristors.

[51] Hong SK, Kim JY, Kim JW, Choi SY, Cho BJ, 2010. Flexible Resistive Switching Memory. Device Based on Graphene Oxide, IEEE Electron Device Lett 31, 1005.

[52] Hong SK, Kim JE, Kim SO, Cho BJ, 2010. Analysis on switching mechanism of graphene oxide resistive memory device, J of Appl Phys 31 1005 (2010).

[53] Deegan R.D., Bakajin O., Dupont T.F., Huber G., Nagel S.R.and Witten T.A., 1997. Capillary flows as the cause of ring stains from dried liquid drops, Nature 1 827.

[54] Deegan R.D., Bakajin O., Dupont T.F., Huber G., Nagel S.R.and Witten T.A., 2000. Contact line deposits in an evaporating drop, Physical Review B, Vol. 62, n°1, p.756.

[55] Supercapacitor electrode based on mixtures of graphite and carbon nanotubes deposited using a new dynamic air-brush deposition technique, P Bondavalli, C.Delfaure, P.Legagneux, D.Pribat JECS 160 (4) A1–A6, 2013.

[56] Carbon Nanotubes based Transistors composed of Single-Walled Carbon Nanotubes mats as Gas Sensors: a Review, P.Bondavalli, Comptes-Rendus de l'Académie des Sciences (Elsevier), September 201.

[57] Resistive Switching in Al/Graphene Oxide/Al Structure, G N. Panin, O O. Kapitanova, S Wuk Lee, A N. Baranov, and T W Kang, Japanese Journal of Applied Physics 50 (2011) 070110.

[58] Vasu K. S., Sampath S., Sood A. K. 2011. Nonvolatile unipolar resistive switching in ultrathin films of graphene and carbon nanotubes Solid State Communications, 151 (16). pp. 1084–1087.

[59] Milaninia, K. M.; Baldo, M. A.; Reina, A.; Kong, J. All Graphene Electromechanical Switch Fabricated by Chemical Vapor Deposition. Appl. Phys. Lett. 2009, 95, 183105.

[60] Kim, S. M.; Song, E. B.; Lee, S.; Seo, S.; Seo, D. H.; Hwang, Y.; Candler, R.; Wang, K. L. Suspended Few-Layer Graphene Beam Electromechanical Switch with Abrupt on-off Characteristics and Minimal Leakage Current. Appl. Phys. Lett. 2011, 99, 023103.

[61] Inside NAND flash memories, Rino Micheloni, Luca Crippa, Alessia Marelli, Springer, e-book, ISBN 978-90-481-9431-5.

[62] Patent US4531203 (A) – Semiconductor memory device and method for manufacturing the same Masuoka F., Iizuka H.

[63] Masuoka, F.; Momodomi, M.; Iwata, Y.; Shirota, R. (1987). "New ultra high density EPROM and flash EEPROM with NAND structure cell". Electron Devices Meeting, 1987 International. IEEE.

[64] Nonvolatile Memory Technologies with Emphasis on Flash: A Comprehensive Guide to Understanding and Using NVM Devices, J. Van Houdt, R. Degraeve, G. Groeseneken, H. E. Maes Published Online: 23 MAR 2007 DOI: 10.1002/9780470181355.ch4.

[65] Angus Kingon, Nature 401, 658–659 (1999), Device physics: Memories are made of ...,

[66] Yuhua Cheng, Chenming Hu (1999). MOSFET classification and operation. MOSFET modeling & BSIM3 user's guide. Springer. p. 13. ISBN 0-7923-8575-6.

[67] http://www.itrs.net/Links/2011ITRS/2011Chapters/2011ERD.pdf

[68] Determination of Work Function of Graphene under a Metal Electrode and Its Role in Contact Resistance, Seung Min Song, Jong Kyung Park, One Jae Sul, and Byung Jin Cho, Nano Letters 2012 12 (8), 3887–3892

[69] Yu, Y.; Zhao, Y.; Ryu, S.; Brus, L. E.; Kim, K. S.; Kim, P. Tuning the Graphene Work Function by Electric Field Effect. Nano Lett. 2009, 9, 3430–3434.

[70] Yan, L.; Punckt, C.; Aksay, I. A.; Mertin, W.; Bacher, G. Local Voltage Drop in a Single Functionalized Graphene Sheet Characterized by Kelvin Probe Force Microscopy. Nano Lett. 2011, 11, 3543–3549.

[71] Haomin Wang, Yihong Wu, Chunxiao Cong, Jingzhi Shang, Ting Yu, Hysteresis of electronic transport in graphene transistors. ACS Nano 11/2010; 4(12):7221–8. DOI:10.1021/nn101950n.

[72] W. Kim, A. Javey, O. Vermesh, Q. Wang, Y. Li, H. Dai, Hysteresis caused by water molecules in carbon nanotube field-effect transistors, Nano Lett. 3 (2003)193–198.).

[73] CNTFET based gas sensors : State of the art and critical review, P.Bondavalli, P.Legagneux and D.Pribat, Sensors and Actuators B, Volume 140, Issue 1, Pages 304–318, 2009.

[74] Echtermeyer T, Lemme M, Baus M, Szafranek B, Geim A, Kurz H. Nonvolatile switching in graphene field-effect devices. Ieee Electron Device Letters. 2008; 29(8): 952–954.

[75] T. J. Echtermeyer, M. C. Lemme, M. Baus, B. N. Szafranek, A. K. Geim, and H. Kurz, A Graphene-Based Electrochemical Switch. arXiv:0712.2026v1.

[76] S. Bertolazzi, D. Krasnozhon and A. Kis Nonvolatile Memory Cells Based on MoS2/Graphene Heterostructures ACS Nano, 2013, 7 (4), pp 3246–3252.

[77] Graphene Flash Memory Augustin J. Hong, Emil B. Song, Hyung Sk Yu, Matthew J. Allen, J Kim, J D. Fowler, Jonathan K. Wassei §, Youngju Park, Yong Wang, Jin Zou, R B. Kaner, B H. Weiller, and L. Wang, ACS Nano, 2011, 5 (10), pp 7812–7817.

Index

About the Editors

Dr. Cristiano Calligaro received the laurea degree in Electronic Engineering and the Ph.D. degree in Electronics and Information Engineering from the University of Pavia (Italy) in 1992 and 1997 respectively. After obtaining the Ph.D. degree he moved to MAPP Technology. In 2006 he established RedCat Devices srl. During his career he was involved in memory design (volatile and non volatile) both for consumer application (multilevel flash memories) and space applications (rad-hard memories) and software design for SEE evaluation using free CAD tools (Open Circuit Design). His current research interest is focused on rad-hard libraries to be used for rad-hard mixed signal ASICs, stand-alone memories (SRAMs and NVMs) and testing methodologies for rad-hard components. He holds about 20 patents mainly in the field of multilevel NVMs and is co-author of about 40 papers. He has been coordinator of RAMSES and ATENA projects inside the Italy-Israel Cooperation Programme, SkyFlash project in the European FP7 Programme and EuroSRAM4Space project in the Eureka Eurostars Programme. In 2014 he co-founded BlueCat Energy as a spin-off company of University of Palermo and in 2017 he co-founded RCD Microelectronics in Hong Kong. He is IEEE member and Eureka Euripides reviewer.

Dr. Umberto Gatti was born in Pavia (Italy) in 1962. He received the Laurea and the Ph.D. degree in Electronics Engineering from the University of Pavia in 1987 and 1992, respectively. From 1993 to 1999, he worked in the R&D Lab of Italtel, as ASIC Designer, being involved in modeling analog ICs. In 1999, he joined the Development Technologies Lab of Siemens, where he was Sr. Design ASIC Engineer. Besides developing telecom ICs, he was the coordinator of Eureka-Medea+ projects focused on high-speed sigma-delta converters. In 2008 he moved to Nokia Siemens Networks where he worked as Sr. Power Supply Architect. Since 2012 he is with RedCat Devices as member of the Executive Staff and also hold co-operation with the University of Pavia. During his career he was involved in the design of data-converters, broadband wireless transceivers, Hall sensors micro-systems and power supply architectures. His current research interests are in the area of rad-hard CMOS ICs, particularly rad-hard libraries and memories (SRAMs and OTP), rad-hard mixed-signal circuits (ADC-DAC) and in testing such components. He holds 2 international patents and is co-author of about 60 papers. He is Senior Member of IEEE and serves also as reviewer for magazines and conferences.

For Product Safety Concerns and Information please contact our EU
representative GPSR@taylorandfrancis.com
Taylor & Francis Verlag GmbH, Kaufingerstraße 24, 80331 München, Germany

www.ingramcontent.com/pod-product-compliance
Ingram Content Group UK Ltd.
Pitfield, Milton Keynes, MK11 3LW, UK
UKHW021110180425
457613UK00001B/22